# Astrological Alchemy

## How to Turn Your Base Transits into Golden Opportunities

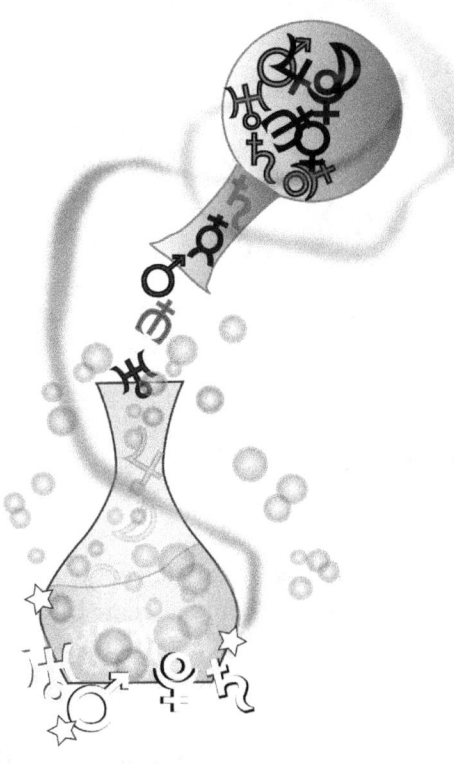

## Joan Negus

Astrological Alchemy:
How to Turn Your Base Transits into Golden Opportunities
Second Edition

© 2010 by The Estate of Joan Negus

All rights reserved. No part of this book may be reproduced or used in any form or by any means—graphic, electronic or mechanical, including photocopying, mimeographing, recording or information storage retrieval systems—without written permission from the publisher. A reviewer may quote brief passages.

by Joan Negus

First Edition, First Printing as
Astro-Alchemy: Making the Most of Your Transits
1985 by ACS Publications

Cover and book design by Maria Kay Simms

Library of Congress Control Number: 2010926951

International Standard Book Number: 978-1-934976-10-4

Published by Starcrafts Publishing
an imprint of Starcrafts LLC
334-A Calef Highway
Epping, NH 03042

Printed in The United States of America

# Contents

Illustrations ........................................................................... vi

## Chapter One
Introduction ............................................................................. 1
    Aspects ............................................................................. 4
    Alchemy ............................................................................ 5
    Categorizing Transits ........................................................ 6

## Chapter Two
Moon, Sun, Mercury, Venus ................................................. 11
    The Moon ........................................................................ 11
    Sun, Mercury, Venus ....................................................... 13
    Lunar and Solar Returns ................................................. 14
    Lunations and Eclipses ................................................... 19

## Chapter Three
Mars ....................................................................................... 22
    Function .......................................................................... 22
    How You Might Feel with Mars Transits ....................... 22
    How Mars Transits May Be Used .................................. 23
    Alchemy .......................................................................... 25
    Transiting Mars in Houses .............................................. 27
    Transiting Mars in Aspect .............................................. 28

## Chapter Four
Jupiter .................................................................................... 29
    Function .......................................................................... 29
    How You Might Feel with Jupiter Transits .................... 29
    How Jupiter Transits May Be Used ............................... 30
    Alchemy .......................................................................... 34
    Transiting Jupiter in Houses ........................................... 35
    Transiting Jupiter in Aspect ............................................ 36

# Chapter Five

Saturn .................................................................................................. 38
    Function ........................................................................................ 38
    How You Might Feel with Saturn Transits ........................... 38
    How Saturn Transits May Be Used ...................................... 39
    Alchemy ....................................................................................... 45
    Transiting Saturn in Houses ................................................. 47
    Transiting Saturn in Aspect .................................................. 48

# Chapter Six

Uranus ................................................................................................. 50
    Function ........................................................................................ 50
    How You Might Feel with Uranus Transits ......................... 50
    How Uranus Transits May Be Used ..................................... 51
    Alchemy ....................................................................................... 59
    Transiting Uranus in Houses ................................................ 60
    Transiting Uranus in Aspect ................................................. 61

# Chapter Seven

Neptune .............................................................................................. 63
    Function ........................................................................................ 63
    How You Might Feel with Neptune Transits ....................... 64
    How Neptune Transits May Be Used ................................... 65
    Alchemy ....................................................................................... 74
    Transiting Neptune in Houses .............................................. 75
    Transiting Neptune in Aspect ............................................... 76

# Chapter Eight

Pluto ..................................................................................................... 78
    Function ........................................................................................ 78
    How You Might Feel with Pluto Transits ............................. 79
    How Pluto Transits May Be Used ......................................... 81
    Alchemy ....................................................................................... 91
    Transiting Pluto in Houses .................................................... 92
    Transiting Pluto in Aspect .................................................... 93

## Chapter Nine

The Moon's Nodes ............................................................. 95
    Function ........................................................................ 95
    How You Might Feel with Nodal Transits ........................... 96
    How Node Aspects May Be Used ...................................... 96
    Alchemy ...................................................................... 103
    Transiting Nodes in Houses ............................................ 104
    Transiting Nodes in Aspect ............................................ 105

## Chapter Ten

Putting It All Together .................................................... 107
    Planetary Cycles and the Adult Stages ............................ 107
    Transiting Planets and Nodes in Houses ........................ 114
    Customizing Transits .................................................... 117
        Sign Emphasis by Sign ........................................... 118
        Sign Emphasis by House ......................................... 120
    Monthly Transit Sheets ................................................. 124
Six-Month Transit Sheets ................................................ 126
Alternatives .................................................................. 129

## Chapter Eleven

Answers to Exercises ..................................................... 135
Answers to Mars Exercises ............................................. 135
Answers to Jupiter Exercises .......................................... 136
Answers to Saturn Exercises .......................................... 136
Answers to Uranus Exercises ......................................... 137
Answers to Neptune Exercises ....................................... 137
Answers to Pluto Exercises ............................................ 138
Answers to Node Exercises ........................................... 138

Bibliography ................................................................. 139
Index ........................................................................... 140

## Illustrations

| Figure | | Page |
|---|---|---|
| Natal Chart | | 119 |
| 10.1 | Work Sheet | 119 |
| 10.2 | Monthly Transit Sheets | 125 |
| 10.3 | Six-Month Transit Sheet | 127 |
| 10.4 | Geocentric 30-degree Graphic Ephemeris | 130 |
| 10.5 | Geocentric 45-degree Graphic Ephemeris | 131 |
| 10.6 | Geocentric 45-degree Graphic Ephemeris | 133 |
| 10.7 | Portion of Calendar Transit Page for September 2010 | 134 |

## Chapter One

# Introduction

In our cause-and-effect-oriented society, both astrologers and non astrologers have **scientifically** tried to prove or disprove the validity of astrology. Studies do indicate that the Sun and Moon physically affect life on Earth. Some research has also been done on the planets.[1] But is it necessary to explain scientifically why astrology works? I think not. What is relevant is that as the planets move and form aspects with natal planets and points, there is a high correlation between their astrological definitions and the tendencies prevalent in our lives during these periods. Our collective experience is the evidence for this.

Often transits are used to explain "why" a crisis occurs at a particular time. This information is supposed to help us accept difficulties that enter our lives. But little had been written about **directing** the transits. The student of astrology is frequently programmed to believe that certain planets and aspects indicate that wonderful things will happen, while other planets and aspects can mean only doom and disaster, and we can do little or nothing about it. So we sit passively and wait to see how predictions

---

1. John Anthony West and Jan Gerhard Toonder, *The Case for Astrology* (New York: Coward-McCann, Inc., 1970  145ff

materialize. We persist in this approach even though predicted times of supposed advancement sometimes manifest themselves as excess and extravagance, while potentially restrictive aspects may be connected with periods of accomplishment and consolidation.

Would it not be much more productive and satisfying somehow to make use of the information provided by the transits? Can we not view transits as informative aids to help us work toward our potential rather than as fatalistically accepted absolutes? And are the planets and points they aspect not signifying segments of ourselves on which we should be working? Transits represent energies coming into our lives and consciousness, and have many ways in which they can and will be expressed. First, one must understand the principles ascribed to the planets; next, recognize the range of their mode of expression – their principles can affect us internally or externally, and their initial impact can be desirable or undesirable. If you are aware that there are choices, you can begin to select among them. The planet or point that the transiting planet is aspecting provides further clues as to types of action that can be taken. One should interpret the transiting planet, and the natal planet or point it is activating as a unit, and relate them to your life circumstances.

Let us look at two examples of this procedure. **Saturn** is indicative of such things as solidification, one's life structure, organization and clarification. It can also be associated with limitations and frustrations. **Neptune** is connected with dissolving, with one's spiritual nature, and with higher evolvement. Or it can represent confusion, deception and illusion.

If transiting Saturn is aspecting natal Neptune, you are most likely being advised that you should be clarifying something that is nebulous in your life, or that you should be putting your spiritual nature to some practical use. You could experience the transit as dissatisfaction with some facet of yourself or your life situation, or as your spiritual development being blocked. Appropriate meanings can be found by examining what is happening in your life that coincides with the symbolism of the planets. Even if the manifestations seem to be negative, consider what you would like to take place, and act upon it. Consciously working toward positive results could alleviate the problem.

On the other hand, if transiting Neptune is aspecting natal Saturn, you should in some way rise above the material, or let go of something in your life structure. Negatively, you could sense that you are being taken advantage of, or feel that your stability or material structure is not inevitable. There is always something constructive that can be done.

In both examples you might face the difficulty directly, or try to express your inner needs in another way. If certain paths seem blocked, you can attempt to push through barriers or take an alternate route. You can also activate the transits internally or externally. In other words, you never have to sit idly by as the universe seems to be doing you in.

Aside from the information that aspects give, planets transiting natal houses also provide relevant data. You will find that issues indicated by the natal house and the planetary themes will begin to emerge as the transiting planet enters that area and will continue until it leaves. This can be corroborated with hindsight. When transiting Pluto left my third house and entered my fourth, I looked back over the twelve years it had been there and noted the transformation that had occurred with my manner of communicating and my relationship with my siblings. When this yielded such significant findings, I checked the transits of the other outer planets as they had moved through the houses in my natal chart. Again there was a high correlation between the symbolism of the transiting planet and activities connected with the area it was transiting at the time. These results were not achieved through conscious effort, and some of what occurred I could have happily done without. Now I consider the message of the transiting planet in terms of the house it is activating and incorporate this in my plan of action so that I can hopefully avoid some of the negativity. Transits contain a wealth of material that can help you make your life more fulfilling.

When some planets activate a natal chart, there seem to be more meaningful occurrences than with other planets. It is usually transits of the **outer** planets (particularly Jupiter, Saturn, Uranus, Neptune, Pluto) that mark major developments in our lives. (The main exceptions to this are solar and lunar returns, eclipses and lunations, which will be discussed later.) This is because they are with us for lengthy periods of time. I once read that aspects of the inner planets are like a neighbor coming over for a cup of coffee, while transits from the outer planets are like a relative coming to live with you. If you do not like the way you are feeling under a Moon aspect, you can wait a few hours and it will go away, but it is difficult to ignore issues that are with you for weeks, months or, in some cases, even years. Aspects from the inner planets can best be incorporated as a shorter segment of a longer course of action that will be indicated by aspects from the outer planets. Mars has a special function. Although it moves relatively quickly, it can also have lasting significance since it is the planet of initiative. The four-day periods in which it forms aspects can denote times to take definitive action within the indications of the slower-moving planets.

# Aspects

Traditionally, trines, sextiles and some conjunctions are considered "good." This is because they are usually manifested without effort. But they can also mean too much of a "good" thing or that the status quo is maintained because there is no impetus to activate them. Squares, oppositions, some conjunctions and some quincunxes are thought of as "bad" in that they often involve obstacles (squares), or motivate changes of direction (oppositions, conjunctions and quincunxes) and we are impelled to act, whether we want to or not. Frequently, in transit, these aspects will involve pressure from others – particularly oppositions and quincunxes – so that one is sometimes forced to take action. Say, for instance, you are experiencing transiting Uranus opposite natal Uranus. This aspect indicates that there is a desire to be free. This need comes from within, but, it might be forced upon you by an unfaithful spouse.

The matter of orbs is a controversial issue. Rarely will an **event** occur on a day on which an aspect is exact. Rather you will experience say a Jupiterian **period** as Jupiter applies to, then separates from, a particular aspect to a natal planet or point. The entire time can be associated with appropriate symbolism at various levels. Jupiter advises you to expand your horizons. Jupiter themes can be materialized through events and/or states of mind. If Jupiter is aspecting your Sun, you might attain ego recognition (external and positive); determine new ego needs (internal and positive); find others are demanding too much of you (external and negative); or feel that your desires are beyond your capabilities (internal and negative). There will undoubtedly be more than one expression of these energies and more than one choice of direction. It is difficult to ascertain exactly when an aspect comes into effective range.

In this book a one-degree orb both applying and separating is used. Admittedly this is somewhat arbitrary. Transits may be significant for a greater orb time—particularly when a planet goes retrograde, moves several degrees away from the aspect but eventually comes back to that point. The first pass is usually the most dramatic because we are faced with a new combination of themes. As we then familiarize ourselves with the possibilities of the transit, and incorporate them into our lives, we can use them beneficially. It is the total acceptance, or denial of transit themes that creates problems. It is true that we are not always able to control what comes into our lives, but we certainly can have some control over our reactions. Sometimes it is difficult to make full use of the information provided by transits because we are apprehensive of change, and cling to even unpleasant situations merely because we are familiar with them. There are, however, ways of utilizing the energies available to us without taking drastic action. I consider the techniques of doing this "alchemical."

## Alchemy

Alchemy was "a medieval form of chemistry, aiming chiefly at discovering methods for transmuting baser metals into gold and finding a universal solvent and elixir of life."[2] It is often overlooked, however, that the alchemist performing rituals was also "purifying his soul."[3] If we then think of transits as opportunities for development and if we perform rituals designed to direct the manifestations of the planetary themes, we can "transmute"—if not into gold or a universal solvent—at least the trends in our lives into more acceptable patterns. These "alchemical" techniques are not ends in themselves, but rather means to ends. They may not even be necessary if we are aware of what we **should** do, and are willing to take action. For instance, transiting Pluto frequently indicates that one should transform oneself in some way. In order to transform yourself, you have to rid yourself of factors that impede that transformation. If this is a frightening prospect, you can begin the process symbolically by cleaning out closets and drawers. This literally makes more space and frequently provides models for action on a more significant level. This may at first seem absurd – but experience has proven that it works!

There are other ways to find clues about the meaning of outer planetary transits, and you should use them all. One way is, of course to learn the definitions given in astrology books. Another is to analyze your own experiences and those of other people during particular transits. Although external circumstances may vary widely, the internal reactions or feelings of most individuals dealing with a particular transit can be quite similar. The basic issues are mirrored in the sky. How we face the issues is up to us. The issues are universal. It is astounding how often a client experiencing a Saturn Return will say, "I have to figure out what I'm going to do when I grow up" even though the particular experience and the answer to the question may be different in each case.

---

2. Laurence Urdang, Editor in Chief, *The Random House Dictionary of the English Language* (New York: Random House 1968

3. Richard Cavendish, Editor, *Man, Myth & Magic* (New York: Marshall Cavendish Corporation, Volume 1 1970).

## Categorizing Transits

The transits of slow-moving planets can be divided into three categories. There are, first, those that everyone experiences at given ages. This category is connected with the cycles of the outer planets.

The second category is based on aspects between Jupiter, Saturn, Uranus, Neptune and Pluto common in the horoscopes of most people born in particular years. Since the outer planets move so slowly such aspects will be in range for months or even one or two years and will appear in the birth charts of individuals born during those periods. Thus shared issues are indicated that must always be dealt with by these persons, but are especially clear when the natal aspects in question are activated by the transiting outer planets. These patterns are more intricate than those of the planetary cycles alone since they involve combining the symbolism of two or more planets. The information yielded from this category is not universally applicable but provides insights into all those born in a given period.

The third category includes aspects of the transiting outer planets to all the personal planets and points. This category customizes the transits to the individual natal chart and should be combined with the other categories.

The first category has been dealt with extensively by psychiatrists and social scientists (although probably unknown to them). Much research had been done on the adult stages or adult cycle of life. The scientific findings relate closely to a combination of the cycles of the outer planets, particularly the conjunction, square and opposition of Jupiter, Saturn, Uranus and Neptune to themselves in the birth chart. Although astrology is never mentioned in the studies that were done, the information provided by them can be very valuable to the astrologer.

Of particular interest was a study conducted at Yale, described in *The Seasons of a Man's Life* by Daniel J. Levinson et al.[4] This study not only deals with life stages that are accurately described by astrological aspects connected with these ages, but it also provides the birth years of the participants so that the aspects depicted in each period can be validated by scanning the ephemeris. Then, since the cycles of the planets are fairly regular (with the exception of Pluto), the definitions given will be applicable to anyone at the particular age. For example, the **Early Adult Transition** begins for everyone between the ages of 17 or 18 and ends at 22 or 23. This coincides with transiting Saturn square natal Saturn (20-23), transiting

---

4. Daniel J. Levinson, et, al, *The Seasons of a Man's Life* (New York: Ballantine Books, a division of Random House, 1978).

Uranus square natal Uranus (19-23), transiting Jupiter opposition natal Jupiter (age 18), and then transiting Jupiter square natal Jupiter (age 21). During this period a person "...must remove the family from the center of his life and begin a process of change that will lead to a new home base for living as a young adult in an adult world.... In the **Early Adult Transition** one must start to give up certain aspects of the pre-adult self and world while internalizing other assets as a groundwork for adult development" (p. 75). In astrology, "structure," which includes "family" and "home base," is a term connected with Saturn, while "change" is a key word for Uranus, and "development" can be ascribed to Jupiter.

According to *The Seasons of a Man's Life*, a person's life is divided into four periods: **Childhood and Adolescence** (3-17), Early Adulthood (22-40), **Middle Adulthood** (45-60) and **Late Adulthood** (65+). There are five-year transitional periods that mark the shift of emphasis from one phase to the next. These are the **Early Adult Transition,** mentioned above (age 17-22), **Mid-life Transition** (40-45) and **Late Adult Transition** (60-65). It is interesting to note that the first two periods of transition coincide with at least four hard aspects of the outer planetary cycles – which is not the case with the interim periods. The **Early Adult Transition,** as mentioned previously, includes Uranus square Uranus, Saturn square Saturn, Jupiter opposition Jupiter and Jupiter square Jupiter. The **Mid-life Transition** contains Uranus opposition Uranus, Neptune square Neptune, Saturn opposition Saturn, and Jupiter opposite, then the square itself.

The ages of the **Late Adult Transition** do not fit so perfectly with the planetary cycles. Personally I would move this period to between ages 57 and 62 because that would be the time of the second Saturn Return as well as Uranus square Uranus and two Jupiter aspects. To corroborate this, it is noted that the participants in the study were between the ages of 35 and 45, so the empirical evidence for the first two transition periods is strong, whereas little evidence for the time frame of the third transition period was given. Had individuals going through the **Late Adult Transition** been interviewed, the age segment might have been shifted.

Levinson's book explains the meaning of the periods in terms that are easy to relate to the appropriate planetary combinations, and describes how the participants experienced and utilized these times in their lives. It supplies a great deal of information that astrologers can incorporate into their study of transits with a view to applying them practically. Some of this material will be referred to when the cycle of each planet is discussed in subsequent chapters.

The second category of transits first came to attention in the early 1970's when, in the course of two weeks, my husband and I received eight frantic phone calls. Those calls were from parents whose children, in one way or another, were drastically changing or thinking of changing their life structure, much to the dismay of the parents. The overt manifestations varied, but the general message was the same. We erected the charts of these individuals, who were all born in 1952 or 1953, and found that all had Neptune conjunct Saturn (potential dissolving of structure) in their horoscopes. By transit, Uranus was conjoining this conjunction; therefore, we have the dissolving (Neptune) of structure (Saturn) suddenly (Uranus). We met with most of these people and discovered that they were all at least considering definite change, although all did not ultimately act upon the urge. These aspects in the natal charts are part of the life pattern and when they are activated by transits of the outer planets, these issues are brought to the foreground. If you are a consulting astrologer, you will notice that during a particular period a number of clients with the same birth years and configurations will come for consultations. This will be when certain transits of the outer planets are in range. If you listen carefully to the first few who come you will know the message of the transits, and, even from the first client, can begin to determine how to make the most of the planetary motifs.

Another way to obtain useful information about aspects common to certain years is to examine what was happening in the world at the time of birth, then apply the principles to the individual. A birth chart, of course, is also a mundane chart for that time in history, and the world atmosphere will be reflected in the native's behavior patterns. An example of this is the Pluto-Saturn conjunction of 1946-1948. This conjunction occurred in Leo, a dramatic and overt sign. Historically this period marked the end of World War II, a time when the world was transforming and power was an issue. One way in which these matters were expressed was in the division of Germany. The country was partitioned under the protection of the Allied forces—the powers, which won the war. Rules and regulations (Saturn) were enforced (Saturn and Pluto) and Germany was transformed (Pluto). For those individuals born between 1946-48 with this conjunction, power and/or transformation will play an important part in their lives and will be particularly notable when the Saturn-Pluto conjunction is being activated by transit. Since the conjunction was in Leo, the power issues in the lives of the individuals born during that period will probably be obvious and openly manifested.

Comparing this to the Saturn-Pluto conjunction of 1981-83 which occurred in Libra, although the same planetary energies were being expressed, the mode of expression in Libra will be quite different from that in Leo. The sign of Libra suggests peace and harmony, and the power issues of that period were less definitive than those of the mid-forties. Historically, protests to existing authorities in the early 1980's took the form of sit-ins and human chains as opposed to forcefully taking charge. Persons with the Libra conjunction of Saturn-Pluto will probably deal more subtly with power and transformation in their personal lives.

In direct contrast to both of these conjunctions were the mid-sixties when Uranus, rather than Saturn, conjoined Pluto. These years were marked with sudden rioting in our cities. The control and order of Saturn was replaced by the erratic and revolutionary tendencies of Uranus. Where the Saturn-Pluto conjunction in any sign is indicative of moving slowly and ceaselessly in one direction, and will be inherent in individuals born with this aspect, people with the Uranus-Pluto conjunction will be more prone to spontaneous, displays of power or sudden transformation.

There are, of course, many such examples that could be cited, but that would be a book in itself. Suffice it to say that examination of world conditions at a person's birth time will provide insights into the basic character of that individual. This can then broaden understanding of what the native needs and experiences during particular transits.

It is the third category of transits, however, that defines the total importance to the individual because it includes the first two categories, and combines them with the personal planets, and points. These transits cannot be ascribed to a particular age or birth year, but are uniquely the individual's. You may share transiting Uranus square natal Uranus with a group of people born within two years of your birth. If your natal Uranus is conjunct Jupiter, you will have transiting Uranus also square natal Jupiter in common with those born within months of your birth. If your natal Uranus-Jupiter conjunction is opposite your Sun or Venus, a dimension is added which is limited to those born within a few days of you. And we narrow the shared experience further when houses are brought into the picture.

Each transit symbolizes general qualities that can be applied to any life, and this knowledge is helpful, but it is only through a natal delineation that transits can be customized for a particular individual. By interpreting the natal horoscope the essence and *modus operandi* of the person can be determined in order to most effectively and beneficially utilize the information offered by the transits. For instance, you would not advise people with strong air emphasis to move quickly ahead without planning strategy. Probably they would not act at all, or if they did, the results would be less than satisfying. Besides understanding the individual, the more you know of the life circumstances at the particular period you are investigating, the more specific the choices and direction become.

Even with all the above information, however, it is important to be aware of both potentially positive and negative ramifications of any transit. The reason for this is that a negative manifestation may occur before you have had an opportunity to develop the positive side; or some unpleasant situation could arise while you appear to be moving in the right direction. Neither case is hopeless. By reexamining your position you might discover that only minor adjustments need be made. Remember transits describe conditions, not outcome, and they symbolize a force that can be utilized to help work toward desired ends.

It would be impossible in a single volume to tailor transits to each person's horoscope. Therefore the natal delineations and specific utilization of the transits will be left to the reader. What **will** be discussed are the general meaning and purpose of the transits on various levels; possible application of their themes to bring out positive qualities; and any other information that will enhance understanding and use of them.

## Chapter Two

☽ ☉ ☿ ♀

# Moon, Sun, Mercury, Venus

Timing is essential in forming an effective "game-plan" for utilizing data supplied by transits. So we shall begin with the daily movement and cycle of each planet, duration of aspects, and then discuss how the messages presented by the transiting planets can be used. The Moon, the Sun, Mercury and Venus are covered in a single chapter because their implications are usually less profound than the other planets. The Sun, Mercury and Venus can often be treated as a unit since they are never far from each other.

## The Moon

The Moon completes a cycle in about 29 days. This means that it moves approximately 12 degrees a day. In about 2 days it traverses 30 degrees, and thus forms an aspect of the 30 degrees (12th harmonic) series with every planet or point in a natal chart. These aspects will be in range for about 5 hours (including one degree applying and one degree separating). Frequently lunar aspects will pass unnoticed because they are so rapid or can be associated with quick mood changes during the course of a day. But singly, they are rarely indicators of monumental events.

Lunar aspects, however, can take on more meaning in periods of crisis since during such times there is a heightening of interest in every detail. It is not necessary to regularly track every Moon transit, but you may want to do this when the emotions are in the foreground or when you wish to initiate a change. The transiting Moon can connect emotions with the issues ascribed to the planet it is aspecting or indicate change or fluctuation in the area represented by the house it is transiting. You might, therefore, want to examine Moon transits on a day when a client is going to court for a divorce hearing; or you could note when the transiting Moon is entering a house and take an action appropriate for that area with regard to your emotions or feelings. Some of the possibilities for each house are as follows:

**Moon in the first house** – do something that will be self-satisfying and emotionally gratifying.

**Moon in the second house**
Initiate a project that could bring in income or bolster your feelings of self-worth.

**Moon in the third house**
Take a short trip, communicate with neighbors or siblings.

**Moon in the fourth house**
Set up a dinner party or some other activity in the home.

**Moon in the fifth house**
Plan a romantic evening, speculate or be involved with your children.

**Moon in the sixth house**
Set out to accomplish a small task that can make you feel good, have a medical checkup.

**Moon in the seventh house**
Bring up emotional matters with a partner, express your feelings to those with whom you are close.

**Moon in the eighth house**
Apply for a loan, enter a joint venture in which you can share resources.

**Moon in the ninth house**
Embark on a long journey, take a course of instruction, go to church.

**Moon in the tenth house**
Since the Moon also symbolizes the public, you could schedule a public presentation that would further your career.

**Moon in the eleventh house**
Join a group, get involved with a cause with which you can be emotionally attuned.

**Moon in the twelfth house**
Delve deeply into your own feelings, start work in an institution that could be emotionally gratifying.

It is true that the Moon plays a greater role in the lives of some people than others. If, for instance, you are strongly Cancerian or have a prominent Moon in your natal chart, Moon phases and aspects will be more obviously significant. Still, one should not become so obsessed with tracking lunar transits, however important they sometimes are, that one loses sight of the total life picture.

# Sun, Mercury, Venus

Single aspects from the transiting Sun to your natal chart are also less critical than the transiting outer planets. The Sun makes a complete circuit of the zodiac in a year. Therefore, it moves between 57 and 62 minutes in a day and is in range of an aspect for about 48 hours. Solar aspects may indicate the bringing of matters to light in terms of the planet or point being aspected, or matters concerning the house it is transiting and/or the house that it rules. Sun aspects sometimes are connected with one's physical energy level. I have a client who had allergies, which not only made her uncomfortable, but also exhausted her. They emerged for a few days each spring and fall, and seemed to correspond with transiting Sun square her Mars-Saturn conjunction. When we discussed alternatives to going to bed with this aspect, she decided that instead she would select those periods to initiate (Mars) something that she wanted to accomplish (Saturn) in order to gain recognition (Sun). For several years now she has done this. She has not become world famous, nor have her symptoms totally disappeared, but she is no longer incapacitated when the aspect is in range.

Geocentrically Mercury and Venus are never far from the Sun. Although the Sun, Mercury and Venus are not always conjunct each other, their proximity frequently allows them to be treated together.

Mercury can move a little over two degrees a day when it is travelling at maximum velocity so that its aspects are often in range for less than two days. Mercury represents communications and/or mental activity. For instance, transiting Mercury aspecting Saturn informs you that you could discuss serious subjects, or put something important in writing. When Mercury is retrograde, it is not the most auspicious time to sign contracts or take strong definitive action, especially with Mercurial matters. It is an excellent period, however, to do research, put details in order and make tentative plans. If you must make decisions under a retrograde Mercury, you can still accomplish your goal, but there may be some adjustments that have to be made. If you hold this attitude, the world does not have to come to a halt during a retrograde Mercury period.

Venus is the planet of pleasure and sociability, and Venus aspects can offer a reason to involve yourself in enjoyable activities. You might not be industrious during these periods and may even be self-indulgent or lazy. But since Venus moves so quickly (a maximum of 76 minutes a day), there is usually not even enough time to feel guilty about these potentially negative manifestations. And we all need a break from the rigors of life!

Make use of Sun, Mercury and Venus aspects but do not dwell on them. If you do, you could miss the message of the period. Since the transiting Sun spotlights a planet or area, transiting Mercury denotes communications and transiting Venus, sociability, you could apply them together in some form of social interaction in terms of the house(s) they are transiting. Or they can be utilized in conjunction with the aspects of the outer planets. Individually, their transits have little long-term significance in your life.

## Lunar and Solar Returns

There are, however, certain techniques based on the transiting Sun and/or Moon, which present a larger picture and represent longer time spans than single aspects from the Sun or Moon. One of these is the lunar return, which is a chart that is calculated each time the transiting Moon conjoins your natal Moon at your place of residence. The lunar return generally describes the overall picture of your life for the period in which it is valid. It can help prepare you for what is going to happen in your life during that month, but it can also give you ideas as to how and where you can bring out particular energies.

The house in which the Moon is posited is highlighted for the month and indicates an area on which you should focus attention during that time.

The energies represented by the planets will be most emphatically expressed through the houses in which they are posited, and you should make use of them. For instance, with the lunar return Saturn in the fourth house, you could wait to see if you are going to be restricted in the home, **or** you could work hard and put things in order at home.

Planets within five degrees of the angles will describe energies that will be prevalent in the house in which they fall, but also qualities that will be all pervasive during the whole month. Let us say you have Uranus in the tenth house of the lunar return. You could change jobs that month or you might have to deal with the unexpected in your career. If that Uranus is conjunct the MC as well, change and spontaneity will be a general theme during that period. You can apply Uranian principles by undertaking a creative project or instituting new exciting activities. The more planets that are near the angles, the more probable that the month will be active and memorable.

Aspects within the return (5 degree orb) and between the return and the natal planets and points (2 degree orb) will also be significant.

A lunar return can be interpreted by itself or delineated within the framework of the solar return. The lunar return sets the **tone** for the **month**, but the solar return sets the **scene for the year.** Or you might say that the solar return describes the entire act of a play, while the lunar returns depict each scene within the act. If available time for interpretation is limited, the solar return should take precedence over the lunar.

A solar return is a chart that is calculated for the moment each year that the transiting Sun exactly conjoins your natal Sun—so it occurs on your birthday, the day before or the day after. There is a difference of opinion as to the locality for which it should be erected. Some astrologers believe it should be set up for the birthplace. Others feel it should be calculated for wherever you are at the moment the return begins. And still others erect it for your place of residence, regardless of where you may be temporarily at the time of the return. I use the last alternative for two reasons. First of all, the manifestation of the chart seems to begin two or three months early, overlapping with the previous return. Therefore your location on the day of the return is irrelevant. And secondly, if you move a distance during the solar year and relocate your solar return, you will find that the emphasis in your life shifts and will be reflected in the relocated solar return.

**The following are factors that should be investigated in order to interpret the solar return:**

- houses in which the planets, nodes and Part of Fortune fall
- nodes and signs on the angles
- planets within 5 degrees of the angles
- aspects and configurations within the solar return
- aspects between solar and natal planets and points
  (using a 2 degree orb)

You can also place the natal planets and points around the solar return and interpret them much as you do the solar return planets according to the houses in which they fall. For instance, if the solar Uranus falls in the seventh house of the solar return and the natal Uranus falls in the fourth, changes or an urge for freedom would be indicated both with partnerships (seventh house) and within the home area (fourth house).

The solar return does not describe the personality, but rather the energies and general types of experiences connected with the specific planets and areas of life depicted by the houses. For example, if your natal Moon is in the sign of Cancer and your solar return Moon is in Capricorn, this does not mean that for this year you are going to be practical and duty-oriented about your emotions. Rather, that you will probably be dealing with practical emotional **issues**. If the solar return Capricorn Moon is in the sixth house of the return, you might find yourself dealing with the emotions of coworkers or subordinates which will need to be placed into some practical perspective. This type of interpretation seems to work quite well for the personal planets, but less so for the outer ones because the latter are in a sign for extended periods of time. This means that everyone will have Neptune in the same sign for 14 or so years in solar returns. This is significant for society-at-large, but may not be as pertinent for the individual.

The particular energies represented by the planets and the houses in which they are posited are, therefore, more important than the signs in which they appear.

The **house in which the Sun falls** (natal and solar Suns will, of course, be conjunct) will be an area of importance in the year. In a natal chart the position of the Sun indicates where the native can "shine." In a solar return it shows where things will come to light or where the ego will be prominent.

The **Moon's houses** are where there may be fluctuation, where one should have dealings with the public or where emotional experiences will occur.

The **Mercury** placements will show where communications will be important or indicate subjects that should be discussed.

With **Venus and Jupiter**, beware of self-indulgence (Venus) and excesses and extravagance (Jupiter); but these energies can be used to enjoy (Venus) and develop (Jupiter) oneself in terms of the areas involved.

The **Mars** positions are where you should take the initiative or assert yourself. There can also be a great deal of activity in these areas.

**Saturn** placements can mean delays or restrictions, but should be used to solidify one's position or clarify situations connected with the houses in which they appear. I had a student who was having marital problems. Her current solar return had Saturn in the seventh house. She called the day she discovered this and said, "I don't want my marriage to end, but isn't that what Saturn in the seventh means?" I agreed that it could, but suggested that it might also be a good time to put the relationship in order. The choice was hers, and her husband's of course. That was the year they went to a marriage counselor and their marriage became stronger.

**Uranus** house positions are areas in which there will be changes or where you may change your attitude. You should try to be flexible, expect the unexpected or create excitement in these areas.

**Neptune** can be self-deception, deception or unclarity, but, should be used to become more highly evolved. There can be the tendency with Neptune to let one's imagination run rampant so you should not allow suspicions to build up in the areas of the Neptune placements. By facing up to unclear situations, you might discover that there was no basis for your feelings; or if you were right, you might remove them by bringing them to the surface.

The house positions of **Pluto** will indicate where there may be transformatory changes; or issues of power (or dealing with power figures) may arise. It is best to probe deeply and analyze any situation that can be connected with these areas. They will not remain hidden, and by instituting investigation you could feel more in control. These will also be areas where you should concentrate on eliminating anything that impedes your progress.

Besides the planets, the house placements of the **Part of Fortune** and the **nodes** are also significant. The area symbolized by the house in which the Part of Fortune is posited is where one should think about wholeness or fulfillment. You will want **personal** fulfillment if it is in the first house. In the seventh you could find fulfillment through a partner, etc. The nodal axis reveals two areas in which association with others will be prominent. You might want to make contact with people who could be beneficial to you in those compartments of your life. If the nodes are in houses three and nine in your solar return, and you were a writer, you could write something (third house) and submit it for publication (ninth house).

The modes of the angles will supply more data. They indicate a manner in which you should operate. **Cardinal** angles advise you that this is an initiating year. **Fixed** angles state that you should solidify or work on matters begun before the start of the solar year. **Mutable** angles inform you that you should finish up something. With mutable angles you may also feel at the disposal of others and may have to consider other people's positions before you can take action. If the angles are mixed, combine the meanings. The Ascendant-Descendant axis will describe the personal portion of your life; the MC-IC axis, interaction with the outer world.

The specific sign on the Ascendant will inform you of qualities you should personally express during the year. The following are some of the possibilities for each sign.

**Aries**—be self-assertive, or take the initiative.

**Taurus**—solidify goals, particularly those associated with personal pleasure.

**Gemini**—be gregarious, as adaptable as possible, and communicate.

**Cancer**—emphasis will be on nurturing, or the home (this could be a year in which the nesting quality will be brought to the foreground)

**Leo**—there will be an urge to attain recognition, a year to stand out and not melt into the crowd.

**Virgo**—a good time to organize, concentrate on details or become involved in some service-oriented project.

**Libra**— the way you relate will be emphasized or close one-to-one relationships will demand attention (there could be something to be resolved between the "I" and the "Thou").

**Scorpio**—analyze and probe deeply into important matters

**Sagittarius**— a desire to travel, an urge for freedom, or broadening yourself in some way.

**Capricorn**—career- or success-orientation, desire to make an impression on the world.

**Aquarius**—a time to express your independence or rid yourself of unnecessary burdens.

**Pisces**—a year to be giving and humble or a time for artistic or spiritual endeavors (if you voluntarily help others you can hopefully avoid being taken advantage of).

There are, of course, many other possibilities, and the natal chart as well as the other facets of the solar return must be incorporated into the total meaning.

Aspects and major configurations within the solar return can describe intricate issues in the year but will be most evident when connected with natal planets and points. This will be obvious with aspects of the transiting outer planets because they will be in range for most, if not all, of a year and therefore will appear in all solar returns for that period. For example, if transiting Uranus were squaring transiting Saturn, this would be in range for a number of weeks and would be in all solar returns calculated for that time. This means that everyone whose birthdays fall during those weeks would be experiencing the freedom (Uranus) versus responsibility (Saturn) dilemma in that year. But if this square formed a T-square with your natal Moon, the matter could become emotional, or there could also be fluctuation between doing your duty and expressing your independence. On the other hand, someone not having this mundane square connected with any planet or point in his/her natal horoscope might not experience the Saturn square Uranus as fully or dramatically.

Planets within five degrees of the angles of the solar return will also be indicative of energies that will be prevalent in one's life during a particular year. It seems to work in a general way but can be applied to the particular angle. For instance, an angular Jupiter would imply a year to expand and develop. In the first house you could connect it with personal growth. In the fourth house you might select that time to build a room onto your house. In seventh house you could choose to get married. In the tenth house you might ask for a raise.

One way or another, the picture painted by the solar return will unfold during the year. It can be thought of as a still life, which you can passively observe to prepare you for what will occur. Or you can view it as a motion picture in which you can participate—a movie of which you can even be producer, director and the lead character.

## Lunations and Eclipses

Other charts which can be valuable in planning strategy are those of lunations and eclipses. Both of these types of charts are also cast for the locality in which you reside. A year usually has thirteen pairs of lunations (erected for the moment the Moon conjuncts or opposes the Sun in the sky) and, these lunations also help to describe the issues prominent to you in a month. Eclipse charts set for New (Sun conjunct Moon) and Full (Sun opposition Moon) Moons are meaningful for longer periods of time. One theory is that solar eclipses are significant for a number of **months** (equivalent to the number of minutes in which the eclipse is total). Lunar eclipses are said to be important for the number of **weeks** that are equivalent to the number of minutes in which the eclipse is total.

It is helpful to place both lunation and eclipse charts around the natal chart. A lunation or eclipse chart erected for your locality will apply to that community. The placement of the planets and points (including the MC and Ascendant) around your horoscope make it more personally yours. In this way you can see the areas emphasized in your natal chart, and the interaction between the lunation of eclipse chart and the natal chart.

The New Moon is a time to initiate situations that will culminate when the Moon is full. Within the course of the month you should see some results of your actions. If you have a small task or project you wish to accomplish, you might organize it within the framework of the lunation chart. For instance, you might be planning a trip to Europe. You could pick the day of the New Moon to either start the journey or to make arrangements for it, especially if the Sun-Moon conjunction were to fall in your ninth house. If at the same time the Saturn of the lunation chart falls in your seventh house, you could anticipate some resistance from your partner and, therefore, to avoid this possibility you should explain every detail or ask your partner to help organize the trip. You can incorporate the other planets in your course of action as well, with special emphasis on interaspects between the lunation chart and your natal chart. A great deal of information and guidance can be gleaned from the lunation chart.

Eclipse charts have more long-term implications. Astrologers accept the premise that eclipses can have a strong impact on the individual, especially when the Sun and Moon form close aspects with natal planets and points. In the past, the attitude toward them has been usually negative. People seemed to expect dire results and to look for manifestations to support this view. Modern astrology tends to be more optimistic, but there is still much negativism about eclipses.

If we can direct simple transits, why not at least try to use the powerful energies ascribed to an eclipse to our advantage? I have an exercise that I do in my classes prior to an eclipse. We place the planets, MC and Ascendant of the eclipse chart around each student's horoscope. Then we create a scenario. Eclipses do not materialize instantaneously at the moment of the eclipse nor disappear as soon as it has passed. It represents a slowly unfolding process, which we can take into consideration in our plans. Feedback has shown that ultimate results are not seen until much later.

In formulating a scheme, the house(s) in the natal chart in which the eclipse Sun and Moon fall are noted first and should be thought of as focal point(s). Think of something you would like to accomplish in terms of the areas involved. Then incorporate the other planets and points in the eclipse. Pay particular attention to **any** close inter aspects between the eclipse planets and the natal planets and points. These, of course, are transits of that particular time but are made even more prominent when they are part of an eclipse chart.

At the time of the eclipse or just before it, initiate some action in terms of the scheme you have created. It is surprising how this little nudge can move matters along. I wrote my first book under a series of eclipses that fell in the third and ninth houses of my natal chart. Each time an eclipse occurred during this period I would do something specific connected with the book and the book appeared in a relatively short period of time. There were a few delays, but the process, for the most part, went quite smoothly.

Thus solar and lunar returns, lunations and eclipses are forms of transits that present an overview for a given time span and can provide ideas for influencing the outcome of some situations. But examining the transiting outer planets singly can be advantageous as well. They can enhance the meanings of the aforementioned charts and also suggest individually appropriate activities to develop facets of character of life-direction during particular periods.

# Chapter Three

## ♂ Mars

The Mars cycle is 1.88 years in length. Its maximum movement is 46 minutes a day, and it is in a sign for about two months. A Mars aspect will be in range for about four days, and even longer if it is retrograde. Although its implications are not as profound as those of the outer planets, Mars transits can be a symbolic trigger for major developments when combined with the information provided by transits of the other outer planets.

### Function

The function represented by Mars transits is to initiate, to direct, to be self-assertive and to be actively involved.

### How You Might Feel with Mars Transits

Because Mars transits last longer than those of the Sun, Moon, Mercury and Venus, they can be associated with more identifiable feelings and experiences. Not that Mars, or any other planet, is responsible for behavior; but as planets become prominent by transit, there is a correlation between their symbolism and manifestations in our lives.

When Mars is in the picture, you may feel energetic, enthusiastic, and have a strong desire to take action or be self-assertive. The energy level may be so high—if it is not expended—that you can become irritable, tense, nervous and impatient. Action may be taken without forethought, leading to situations that are less than desirable, or one can be careless. Self-assertion, if it is thwarted or not purposefully directed, can lead to argumentativeness and aggression.

## How Mars Transits May Be Used

Clues as to how the energies represented by Mars can be beneficially applied may be found in the planets and points it is aspecting in the natal chart, and in the house it is transiting.

With the Mars cycle itself, you can be certain that the days on which Mars aspects the horoscope will not be tranquil. You should select such times to accomplish something that requires forceful action. You might move the furniture or wash windows, or initiate some project, physical or not. Mars conjunct Mars is the best time to start a task that might take two years to come to fruition. In that way you can make use of the entire cycle and reevaluate your position with the squares and opposition. It may be easier to control the energy during the sextiles and trines, but these aspects may lack the impetus for accomplishment that the conjunction, squares and opposition imply.

It is not absolutely necessary to formulate a two-year plan with the Mars cycle. It is merely mentioned in case a two-year plan is in the offing as the conjunction is about to occur. The aspects of the Mars cycle can also be considered singly. These are not periods to plan restful retreats since they will most likely be disrupted in some way, but can be auspicious for making strides forward.

Transiting Mars aspecting the natal **Sun** advises you to assert your ego, or to take the initiative to attain ego gratification. You might be volatile or anger easily during these periods if you are not consciously trying to use this energy in some other way.

When Mars aspects natal **Moon**, you should express your emotions. Your feelings can easily be shown and if you try to repress them there could be emotional outbursts. It is better to consider this a period to get things out of your system. Determine what you want to bring out and choose the time and place to do it.

Mars to **Mercury** aspects can signify an urge to communicate. You may speak more loudly or more quickly than usual. You should, however, try to think before you speak or you might find yourself suffering from "foot-in-mouth disease." If you are writing something, you could be impatient with the project and do a slovenly job, or find that you are leaving out words or thoughts because you are in such a hurry to finish. Here it is advisable to write for short periods of time.

When Mars aspects natal **Venus** you could decide to socialize, express affection or do artwork. If there is anything you wish to accomplish where charm can be an asset, you might select these times to display your charm or sex appeal. On the other hand, you could appear overaggressive or pushy with the opposite sex.

Mars to **Jupiter** aspects are excellent times to create grand plans, or to take the initiative in order to develop your self. Enthusiasm can run very high during these aspects so be cautious not to undertake more than you can handle, monetarily or otherwise.

Mars to **Saturn** aspects indicate times to initiate in order to attain rewards for which you have worked, but will meet with frustration if you do not deserve them. You might also take on some responsibility during these aspects. Select your own tasks or commitments, if possible, to avoid feeling overburdened.

Mars to **Uranus** is a highly energetic combination when either of these planets is transiting the other. There is one quality they share: the tendency to be careless. Natal Uranus craves excitement and transiting Mars "fans the fire," or natal Mars wants action and transiting Uranus says "do it yesterday." In either case, you can be so bent on taking action that you might trip over your own feet or bump into walls. Concentrate on short-term action for which you can see quick results and keep your mind on the task at hand. Mars to Uranus can also indicate times to be creative, and periods in which you should express your individuality.

Mars to **Neptune** aspects can be periods in which you turn dreams or aspirations into reality, or enrich your spiritual life. If you have felt taken advantage of or deceived, as transiting Mars enters the scene you should bring such matters into the open.

Mars to **Pluto** aspects can signify violence, but eruptions usually occur if you ignore the message of these aspects. This combination indicates that you should be taking action toward transformation or asserting your own power. The urge to wield your own power may be strong, and you could be drawn to individuals who challenge this facet of your character or attempt to usurp it. Recognize the possibilities; and instead of either suppressing them or jumping in headfirst, analyze situations thoroughly so that you can be assertive in a controlled manner and not lash out irrationally.

When the **nodes** are involved, interaction with others is connected with the transit. Mars to the nodes could mean that you should take the initiative in a relationship and you will probably not wait for someone else to make the first move. If you just sit back and do nothing, you might find yourself being impatient with others and possibly appearing argumentative.

Since the **Part of Fortune** in the natal chart is indicative of wholeness and of fulfillment, aspects from any transiting planet will affect that concept within yourself. With Mars, you should take some action, which will make you feel better about who you are or what you are doing.

Transits to the **Ascendant and MC** are highly significant. Mars aspecting either of these points can denote high energy, impatience, and/or the desire to be active or aggressive. Since the MC is the cusp of the tenth house, you could use the energy to apply yourself in your career or take the initiative in some other role you perform out in the world. With the Ascendant, you might do something to improve your appearance or be personally assertive.

As Mars transits a house, you may notice that activities pick up in that area. You could become very busy and either wish, or find it necessary, to take the initiative or make decisions in terms of that compartment of your life. Mars will move through a house in about two months; so you can note as it transits each house, determine what you would like to do in that area, and at some point take action.

Rather than going through lengthy definitions of the various possibilities for each house, there is a quiz at the end of this chapter offering some suggestions. There is also a quiz on transiting Mars aspecting planets and points. The purpose of these exercises is to help you expand your understanding of transits. Such quizzes will be found at the end of each of the following chapters on the planets and the nodes. These exercises have been helpful in transit courses. Students seem to grasp the concept of transits readily, and quickly begin to use them creatively. We always discuss possibilities other than those presented and consider them within the frame of reference of natal charts.

# Alchemy

Not everyone is basically assertive. Although we all have Mars in our horoscopes, we must look to the natal Mars by sign, house and aspect to determine how we can best use its transits. Individuals who are not accustomed to overtly asserting themselves may find Mars transits very trying. Mars needs an outlet. If one attempts to hold in Mars energy, one could become nervous or excessively angry. If the energy symbolized by Mars is repressed, one could become ill or prone to accidents. People who have no difficulty taking action can find Mars periods stimulating. They may, however, be over stimulated and should be aware of the possible careless and irresponsible tendencies that can be associated with Mars.

Alchemically, Mars energy can be expended through physical activity. If you feel anger welling up, you might take a walk around the block. If you are still angry when you return, then perhaps it should be expressed; but you will have had time at least to think about what you are going to do or say, and thus possibly avoid doing harm.

If there is a great deal of potential anger in the natal chart, you might want to do something more energetic than just walking. I have one client who vigorously scrubs floors on her hands and knees when she feels she is going into a rage. She finds this very calming for her. She also has the cleanest floors in town! I have another client who keeps a punching bag just for such occasions.

People who have a strong Mars in their charts should probably include physical exercise as part of their daily routine to avoid overload during Mars transits. A young man with Mars conjunct Moon in his natal chart was prone to emotional outbursts. He always felt remorse afterwards and wanted to know what he could do about them. He said that his tantrums were sporadic—in fact, they seemed seasonal. He felt much happier and more productive during the periods in which his temper was under control and wanted to know if he had alternatives to this form of behavior. As we discussed the situation, it turned out that he played baseball in college. During training and baseball season he was much calmer than otherwise. I explained that this was because he had found a suitable and acceptable outlet for expressing his feelings. That fall he went out for football as well, and he now insists that he has become a much more serene individual.

If self-assertiveness is a problem, you might take a course on this subject, especially if Mars is aspecting your Sun or transiting your first house. You will still not become extremely aggressive if it is not in your horoscope, but you could learn techniques that would help you to stand up for your rights and yet be in keeping with your innate disposition.

The carelessness syndrome can be combated by the aforementioned physical exercise, and also by keeping tasks short. If your mind begins to wander while you are doing something, change your activity.

In short, Mars in transit is indicative of high energy, which needs to be directed. Assert yourself and channel your actions in order to move ahead.

## Transiting Mars in Houses

a. Mars in the 1st house
b. Mars in the 2nd house
c. Mars in the 3rd house
d. Mars in the 4th house
e. Mars in the 5th house
f. Mars in the 6th house
g. Mars in the 7th house
h. Mars in the 8th house
i. Mars in the 9th house
j. Mars in the 10th house
k. Mars in the 11th house
l. Mars in the 12th house

1._____Boss your family around, or direct activities in the home.

2._____Argue about religion, or take a long trip on which you can be physically energetic.

3._____Suppress your energy, or probe energetically into hidden motivations.

4._____Spend your money recklessly, or be assertive about your pleasures.

5._____Nag your children, speculate, or take up a hobby, which involves physical activity.

6._____Be overaggressive or do something that will improve your appearance.

7._____Fight with your boss or put your energies into your career.

8._____Argue with your partner or share a physical activity with him or her.

9._____Participate in shouting matches, or state your ideas emphatically.

10._____Bully coworkers or add tasks to your daily routine, especially those that require a lot of energy.

11._____ Push your peers around, run for office in an organization, or direct activities in your social group.

12._____ Tell someone else how to spend his or her money or take the initiative in sex.

# Transiting Mars in Aspect

The following exercise includes possible uses of transiting Mars themes as it aspects planets and points in the natal chart.

a. Mars-Sun  
b. Mars-Moon  
c. Mars-Mercury  
d. Mars-Venus  
e. Mars-Mars  
f. Mars-Jupiter  
g. Mars-Saturn  
h. Mars-Uranus  
i. Mars-Neptune  
j. Mars-Pluto  
k. Mars-MC  
l. Mars-Ascendant  
m. Mars-Part of Fortune  
n. Mars-nodes  

1. _____ Overextend yourself physically or use physical exercise to improve your appearance.

2. _____ Viciously express anger or initiate something for ego gratification.

3. _____ Have emotional outbursts or express your emotions.

4. _____ Think before you act because you may be careless, or do something original or creative.

5. _____ Push other people around or take the lead in relationships.

6. _____ Take action toward personal fulfillment, but being too aggressive may lead you away from your goal.

7. _____ Communicate energetically but try not to put your foot in your mouth.

8. _____ Be argumentative with authority figures or take the initiative in your career.

9. _____ Exaggerate or explain your philosophical ideas.

10. _____ You could spread your energies in too many directions or energetically start one project.

11. _____ You could tire easily or use your energy to turn dreams into reality.

12. _____ Overindulge in social pleasures or be charmingly aggressive.

13. _____ You can initiate powerfully, but proceed cautiously and avoid violence.

14. _____ You could feel frustrated and limited, or plan your actions carefully and produce results.

## Chapter Four

## ♃

# Jupiter

The Jupiter cycle is completed in 11.82 years. It is in a sign for about 1 year and moves a maximum of 14 minutes of arc a day. Therefore a Jupiter aspect is usually in range for about 10 days.

## Function

The function represented by Jupiter is that of growth, development and expansion of consciousness. It also can signify fun coming into your life.

## How You Might Feel with Jupiter Transits

Jupiter is associated with a feeling of well-being and an optimistic attitude. You want to explore new territory. There can be an urge to travel; or you may become dissatisfied with established activities and old ways. New doors may seem to be opening for you. You may wish to seek pleasure and freedom.

Jupiter is considered the "great benefic." There is, however, another side to this planet of abundance. Too many opportunities may come your way and decisions may not be easy to make. You could be bored if not enough is happening. It can be difficult to satisfy Jupiterian appetites; so there is the potential of over indulging yourself. This could lead to excesses and extravagance. Willpower takes a back seat during Jupiterian periods. You also may feel that you are being carried along by a strong current, and that you are not totally in control.

## How Jupiter Transits May Be Used

The natal house containing Jupiter indicates where you need to be in a continual state of development. With transiting Jupiter, the principle remains the same. As it aspects natal planets and points, you are informed that you should expand that facet of character for the time span in which Jupiter is in range.

The Jupiter cycle can be indicative of a twelve-year growth period; so there can be several such cycles in a lifetime. The conjunction will mark the beginning of each one, and the squares and opposition, which occur about every three years, are times to evaluate progress and make adjustments. This can be corroborated by the psychological adult stages.

In *The Seasons of a Man's Life* these stages are described as a series of ladders. Goals are assigned to each "ladder," and certain tasks must be performed to accomplish your goal. Although there is constant movement, the main periods are relatively stable because you are moving in one direction—up the ladder. The movement from one ladder to another is marked by a transition period. In transition periods there are fluctuations and potential crises since you are going to a new ladder and there may be more than one choice of direction.

The common thread throughout the book is the theme of development—a Jupiter principle. If we relate the hard aspects of the Jupiter cycles to the ladders, the conjunctions should fall in the midst of a main period (the beginning), and the squares and oppositions should coincide with transition periods. And this is the way it is. The first transitional period, the **Early Adult Transition**, occurs between the ages of 17 and 22 and contains Jupiter opposition Jupiter and Jupiter square Jupiter. The **Age Thirty Transition** (28-33) also has the opposition and the square and the **Mid-life Transition** (40-45), the square and then the opposition.

According to Levinson, the main periods are: **Entering the Adult World** (22-28) which has the second Jupiter conjunct Jupiter (around age 24), **Settling Down** (33-40) which contains the third Jupiter return (about age 36) and **Entering Middle Adulthood** (45-50) which includes the fourth Jupiter return (around age 48).

Of particular interest is the period of **Settling Down** because it describes a very definite Jupiter theme. At this time "a man builds a second structure and reaches the culmination of early adulthood" (p. 56). An important issue of this period is that of the mentor —the older person (but not a great deal older) who helps the individual to develop. This certainly is a Jupiterian concept. According to Levinson, "At the end of the Settling Down period, from about age 36 to 40, there is a distinctive phase that we call **Becoming One's Own Man**" (p. 60). Up to this point there is usually a dependence on the mentor, and attempting to free oneself of the mentor is definitely the onset of a new phase. Age 36 is the time of a Jupiter conjunction.

I am not saying that you should plan and accept your life in 12-year segments. Rather, I am suggesting that Jupiter conjunct Jupiter can be a reference point for examining one's development, and the entire Jupiter cycle can be used as a yardstick by which you can measure personal progress and to make adjustments. But Jupiter can be used in connection with any natal planet or point. You should understand the function of transiting Jupiter, recognize its pitfalls and apply its positive principles as it forms aspects.

Transiting Jupiter aspecting natal Sun or Ascendant has certain similarities. In both cases it **can** signify feelings of personal well-being. There is a strong desire to develop one's self. You may have to be wary of overeating because Jupiter aspecting either the Sun or Ascendant can be associated with physical expansion. If you need to gain weight, you could easily accomplish that task during these aspects. The main difference between Jupiter aspecting the Sun and Jupiter aspecting the Ascendant is that the Sun indicates ego-involvement. You will want your advancement to be seen. Whereas with the Ascendant your progress needs only to be felt although it may be witnessed. The aspects to the Sun will be in range for about 10 days, during which you should project your ego into the world. The aspects to the Ascendant are also in range for the same period, but if Jupiter is **conjoining** the Ascendant it will then move into the first house and you will have about a year in which you can concentrate on personal development.

During Jupiter to **Moon** aspects, feelings will run very high. These are periods when you can be emotionally demonstrative, and if you do not ordinarily behave in this manner, you may surprise yourself. If you have feelings that you would like to express, you might select the times of these aspects to project them. For those who tend to be overemotional, be wary of allowing your feelings to take over totally. There are similarities between transiting Mars aspecting natal Moon and transiting Jupiter aspecting natal Moon in that they can both indicate excess emotionality. But feelings are focused and direct with Mars while they are more diffuse or they tend to overflow with Jupiter. Anger and irritability are usually connected with Mars, whereas uncontrollable tears would most likely fall in the realm of Jupiter.

Jupiter to **Mercury** can be associated with eloquent verbal communications, or the ability to write profusely. It also, however, offers a time of candor or tactlessness (another "foot-in-mouth" aspect). Anything put into writing may require editing later. Just plan in advance what you wish to say and it will flow. Do not avoid writing because excellent ideas can come during these periods. So what if it has to be edited?

The Jupiter to **Venus** aspects are a combination for socializing, having fun and indulging yourself. The warning here is overindulgence, especially monetarily. You might want to lock up your credit cards and bank books at these times! Or better yet, choose enjoyable activities that are not costly.

With Jupiter to **Mars** you may feel a strong surge of energy. There will be a desire to take action. Profitable plans can be instituted if they have been well thought out. There can, however, be the tendency to overextend yourself or spread yourself too thin with these aspects. Try to concentrate on one activity at a time and be realistic about what you can attain. Otherwise you may expend a great deal of energy and accomplish very little.

Jupiter to **Saturn** indicates that you should extend yourself beyond your present life structure. This could be done by acquiring knowledge that could bring you success and broaden your base of operation. Or you might do something to enhance your material world, such as adding a room to your house. With both Jupiter to Saturn and Saturn to Jupiter there is a need to balance expansion and limitation (or consolidation), but with Jupiter as the transiting planet the emphasis is on "reaching out."

Jupiter to **Uranus** can be associated with creative urges and/or expression of individualism. You will want to express your independence, or you may not feel in total control of your life. You might have to cope with unusual situations that materialize suddenly. You should be as spontaneous as possible during these times and try to "go with the flow." You could possibly avoid some of the unpleasant surprises by creating your own excitement.

During Jupiter to **Neptune** aspects you may find that you are having many dreams, in both a sleeping and waking state. Objectivity and practicality are frequently absent during these periods. There could be marked discrepancies between events and your interpretation of them.

Drugs and alcohol can further distort reality. You might have a reaction even to normally taken medication, or find that a glass of wine affects you more strongly than usual. When these aspects are in range, it is not easy to acquire material gains, but your spiritual life can flourish. You may make dreams come true if they are not totally beyond the realm of possibility. The best method of reaching goals is through visualization, meditation, or other indirect means, and definitely not through aggressiveness.

Since Jupiter represents abundance and free-flowing energy, as it activates **Pluto** you might easily assert your power. Or you could become tyrannical, overzealous and obsessive. But Jupiter also corresponds with the broadening of perspective. If you probe deeply (Pluto) during these aspects, you might discover factors that have been blocking your development, change them and transform yourself in some profound way.

You could find relationships blossoming when transiting Jupiter connects with the **nodes.** You should seek out individuals who can expand your knowledge or whose company you enjoy and avoid those who take up a great deal of your time and offer little in return.

With Jupiter transiting natal **Part of Fortune** you might feel self-satisfaction, or find a way to improve your life direction. If you have been unhappy with your present situation, these aspects can be used to form a new plan to attain fulfillment.

A Jupiter to **MC** transit is the time for interaction with the world. Make your presence known. Perhaps you could undertake something new in your career. Try, however, to be realistic about what you can accomplish so that you can avoid being overwhelmed. One man started a second business during a Jupiter-MC aspect. The opportunity presented itself and it sounded so promising that he did not take the time to investigate it thoroughly. He soon found that there were not enough hours in the day to handle both careers. He ultimately hired someone to run the new firm and it devoured his potential profits. Had he examined all the ramifications before he invested in the second business, he could have saved himself annoyance, exhaustion and money.

As Jupiter transits a house, you will have about a year to focus on development and expansion in matters pertaining to that area. Or, there could be insatiable appetites—wanting more and more in terms of the house involved. In either case, this transit can signify a lack of control —that things are progressing (or failing to progress) without much input from you. Select some way that you would like to improve in terms of that house and take steps toward your goal. You might still feel swept along, but at least you will have chosen the direction in which you are moving.

# Alchemy

Two alchemical applications of Jupiter are to travel and to engage in learning projects. No matter what is being aspected in the natal chart, you can usually find a trip or a course of study that will represent the principles involved. For instance, if transiting Jupiter is aspecting your Sun, you can embark on a journey or take a class that could bring ego-gratification, or both. Making use of Jupiter in these ways can produce results on two levels.

Travel or courses can be a substitute for possible excesses and over-indulgence. Overeating, for instance, often occurs when one is bored or dissatisfied. Trips and courses have a broadening effect that can replace gaining weight. A client, who is a college student, has natal Jupiter conjunct her Ascendant and recently told me that during the school year she never gains weight, but in the summer, weight increase becomes a problem. My advice was to take a class during the summer or to travel, whichever was more expedient.

These same activities that can be used to alleviate or reduce negative manifestations, can also produce direct and overt positive results. Travel may keep you from being excessive in some way, but also is a way to learn a great deal. You might come into contact with people who can broaden your thinking; or be exposed to new customs or cultures that can enrich your life. Courses can expand your knowledge generally, but also may lead to advancement in your career. And attaining any academic degree can open doors in our society.

Do not, however, ignore the fun and freedom that Jupiter often urges you to experience. A trip or course can include both education **and** pleasure. On a journey you can visit historic sights and still find time for social interaction. A rigorous schedule may be inappropriate here, whether you are traveling or studying. You will probably wish to set your own pace. But a difficult course can be approached with the thought of making it fun, or can be balanced with a class taken for sheer enjoyment or one in which you can easily express your own ideas.

If you find yourself bored during Jupiter aspects, take on some new activity that is different from anything else you have been doing. Or take a little time off just to enjoy yourself.

Transiting Jupiter will always have a message that can be used to help you develop yourself. If you feel overwhelmed and swept along by a powerful current that is pulling you in a potentially disastrous direction, you have not found the best possible meaning of the transit. Take time to examine all facets of your present situation and place them in perspective to determine the advice that Jupiter is offering. To give a dramatic example of this: about three years ago a woman came for a reading during a period in which she was having a series of Jupiter aspects. Prior to our session she and her husband had filed for bankruptcy in the family business and there was a serious fire in their home. She naturally felt devastated by the events, could think of nothing else, and wanted desperately to find out "Why me?" In the course of our conversation some very interesting facts came to light. She and her husband had been contemplating a move to another part of the country. Their business and their home had kept them from doing this. Admittedly the manifestation was drastic, but the indications were clear. The obstacles to the relocation had been removed. This new perspective changed her attitude toward their situation. They did move, and their lives took a turn for the better.

Fortunately, although the symbolism fits, Jupiter aspects are not usually so severe. But this example makes the point that it is beneficial to examine the message of transiting Jupiter carefully, and not to assume that all will automatically be well. The following exercises present more common possibilities of transiting Jupiter in aspect to natal planets and points and in its movement through the houses.

## Transiting Jupiter in Houses

a. Jupiter in the 1st house
b. Jupiter in the 2nd house
c. Jupiter in the 3rd house
d. Jupiter in the 4th house
e. Jupiter in the 5th house
f. Jupiter in the 6th house
g. Jupiter in the 7th house
h. Jupiter in the 8th house
i. Jupiter in the 9th house
j. Jupiter in the 10th house
k. Jupiter in the 11th house
l. Jupiter in the 12th house

1. _____ Be obnoxiously overzealous in expressing your views on religion or politics, plan or take a long trip, or enroll in a college course.

2. _____ Become a bigamist, get married, or do something with your partner that is fun.

3. _____ Gamble recklessly, indulge your children, or start an artistic project.

4. _____ Overindulge yourself or work on self-improvement.

5. _____ Scatter your professional activities too much, expand the potentials of your career, then ask for a raise.

6. _____ Become promiscuous, offer to handle someone else's money, share your resources or talents with another individual, or buy a lottery ticket.

7. _____ Party too much or join a group that can contribute to your social development.

8. _____ Overextend yourself financially or spend your money or an activity that can help to broaden you, or launch a money-making venture.

9. _____ Be extravagant in entertaining at home or make reasonable expenditures on home improvement.

10. _____ Deprive yourself of pleasure, start studying the occult, or volunteer your services in an institution.

11. _____ Ramble on in conversation and refuse to listen to others or write letters or a story.

12. _____ Be lazy instead of working, eat lots of wholesome foods to keep you healthy, or add pleasurable activities to your daily routine.

## Transiting Jupiter in Aspect

The following exercise includes possible uses of transiting Jupiter themes as Jupiter aspects planets and points in the natal chart.

a. Jupiter-Sun
b. Jupiter-Moon
c. Jupiter-Mercury
d. Jupiter-Venus
e. Jupiter-Mars
f. Jupiter-Jupiter
g. Jupiter-Saturn
h. Jupiter-Uranus
i. Jupiter-Neptune
j. Jupiter-Pluto
k. Jupiter-MC
l. Jupiter-Ascendant
m. Jupiter-Part of Fortune
n. Jupiter-nodes

1. _____ Overextend yourself for others, go out and meet new people, or work on developing the relationships you already have.

2. _____ Waste your energy or take the initiative on some plan for development and growth.

3. _____ Become overwhelmed by the idea of progress or consider plans for expansion.

4. _____ Spend too much money or enjoy social situations.

5. _____ Allow your need for security to hold you back or extend yourself beyond your normal limitations.

6. _____ Get carried away by self-importance or do something that will make you feel good about yourself.

7. _____ Talk too much, write letters, stories, etc., or communicate with others verbally.

8. _____ Just act strangely or express your individuality and originality to further your development.

9. _____ Do anything that will make people notice you or begin a project that will give you ego gratification.

10. _____ Engage in self-delusion, or work on dreams coming true.

11. _____ Let your emotions rule your life, or openly express them.

12. _____ Ask for a raise that you do not deserve or start a plan for advancement in your career.

13. _____ Become obsessed with introspective thoughts or delve deeply into yourself in order to be transformed.

14. _____ Concentrate solely on having fun and forget fulfillment or move in a direction that will bring you a sense of wholeness.

## Chapter Five

♄

# Saturn

Saturn makes, on the average, a complete circuit of the zodiac in 29.46 years. It is in a sign for about 2-1/2 years and moves a maximum of 8 minutes a day. An aspect is usually in range for about a month. Since Saturn aspects are sometimes slow to materialize, however, one might add a few days to the end of this period.

## Function

The function indicated by Saturn is to clarify, solidify, organize and structure. It is also the planet connected with hard work in order to attain material rewards or some other form of measurable success.

## How You Might Feel with Saturn Transits

Traditionally Saturn is called the "great malefic," and because of this "bad press" is often defined as limiting, confining and depressing. It is true that during transiting Saturn aspects you could feel as though a dark cloud were hanging over you. You might be overburdened with responsibilities or work, restricted by your own limitations, or frustrated in attempts to take action—any or all of which could be depressing. And one should be aware of these possibilities.

But just as Jupiter, the "great benefic," has a negative side, the supposed "great malefic" has a positive one. Individuals experiencing Saturn transits are frequently concerned with success, life purpose and commitment. There is a desire to accomplishment that is tangibly meaningful. With these aspects you want to organize and categorize. You are more apt to view situations objectively and will tend to evaluate in terms of black and white rather than in shades of gray. There is thus an opportunity to clarify issues that previously have been vague. You want to be thorough. If you therefore accept the fact that Saturn aspects correlate with a slowdown of action, you can exercise the required patience.

## How Saturn Transits May Be Used

Success can be attained under Saturn, but it is the kind entailing a slow rise to power that tends to be enduring, rather than a meteoric ascent that is over in a flash. You must apply yourself and attend to painstaking details. When "paying your dues" is the issue, almost assuredly Saturn will be in the picture.

Since Saturn is connected with the material side of our lives, its cycle is probably the easiest to isolate and comprehend. It makes a hard aspect to itself about every 7+ years, and during those times every individual will, in one way or another, examine his/her position in the physical world. The focus may be on the home situation, marriage, career, social status, etc., but always on some very tangible facet of our lives.

The cycle forms the backbone of the psychological adult stages. Although the Jupiterian theme of development is implicit in all the stages, periodic examination and evaluation of one's material world (a Saturn principle) is needed to grow. You must ascertain where you are, and balance this with where you want to go. In order to succeed you have to build on what has gone before and be realistic about what you can achieve. If you were raised in a New York ghetto, the possibility of your becoming king or queen of England is highly unlikely, but there are many stories of individuals who rise from humble beginnings to achieve great success.

Let us compare the hard aspects of the Saturn cycle to the Saturnian issues discussed in *Seasons of a Man's Life*. Naturally, the **transition** periods coincide with hard cyclical aspects of Saturn because before you can move ahead you have to become aware of your present life structure.

The first transition period mentioned in the book is **Early Adult Transition**, which occurs between the ages of 17 to 22. Everyone experiences Saturn square Saturn somewhere between the ages of 20 and 23. The purpose of the period is to make the final move from adolescence to adulthood. According to Levinson, there are two main tasks to be accomplished, and the facets of these tasks which can be ascribed to Saturn are as follows: 1. "…to question the nature of that world (the pre-adult world) and one's place in it; to modify or terminate existing relationships with important persons, groups and institutions; to reappraise and modify the self that formed in it." 2."… to consolidate an initial adult identity…." (p. 56). The specific details will vary according to one's natal chart and early physical environment, but the principles remain the same.

The **Age Thirty Transition** contains the first Saturn Return (ages 28-30). "This transition, which extends from roughly 28 to 33, provides the opportunity to work on the flaws and limitations of the first adult life structure, and to create the basis for a more satisfactory structure with which to complete the era of early adulthood. At about 28 the provisional quality of the twenties is ending and life is becoming more serious, more 'for real'" (p. 58). Could an astrologer have better described the Saturn Return? Furthermore: "…the life structure is always different at the end of the Age Thirty Transition than it was at the beginning" (p. 58). During this period you can reaffirm your commitments or change them, but you will most certainly examine them. This transition period is the doorway to **Settling Down: Building a Second Life Structure**, and the first Saturn Return astrologically is considered the end of the beginning and the start of the second phase of life.

The next Saturn square Saturn which occurs between ages 35 and 38 is the only hard aspect within the Saturn cycle that does not coincide with a transition period that had been experienced by participants in the study. But it is described in the sub-period of **Becoming One's Own Man** (and, we may add, **Woman**) which was mentioned in the last chapter. The Jupiterian mentor is the central figure of the period, but the Saturn task is to rid one's self of the mentor in order "…to become a senior member in one's world, …and to have a great measure of authority" (p. 60)

The **Mid-life Transition** is placed between the ages of 40 and 45, and of all the transitional periods is the most complex and highly publicized. Astrologically this is because it contains Neptune square Neptune as well as aspects within the Jupiter, Uranus and Saturn cycles. This segment of life is sometimes labeled "second adolescence," which is interesting in light of the Saturn cycle because during this time transiting Saturn makes the second opposition to itself (ages 42-46) in the natal chart. The first opposition occurs between 13 and 15—the age of puberty when one is moving from childhood to adolescence and later to adulthood.

The issues of this period are numerous and diversified as one would expect with the symbolism of Jupiter, Saturn, Uranus and Neptune all involved. The Saturnian task which the individual should work on in the **Midlife Transition** "…is to terminate the era of early adulthood. He has to review his life in this era and reappraise what he has done with it" (p. 191).

The thoroughness of *Seasons of a Man' Life* ends here because 45 was the upper age of participants in the study and, therefore, all that was observable. But a tentative view of subsequent periods is offered. This is the point, however, at which I would question the time spans presented. The concepts involved in the later periods, however, have already been clearly defined in the earlier ones, and the astrologer can look to the hard aspects of the Saturn cycle for the time frame of the periods. One can determine the meaning of each stage by reviewing what occurred the last time the same aspect appeared. For instance, the second Saturn square Saturn of the Saturn cycle (ages 50-52) can be compared to the Saturn square Saturn between the ages of 35 and 38; and the second Saturn Return will mark the end of a phase of life as did the first Saturn Return. The specific details will not be the same but the principles will be similar. You need only adjust the latter to present circumstances. Thus, examination of the Saturn cycle can provide a possible backdrop by which everyone can measure one's material development. The Jupiter cycle can be used to examine the overall growth pattern, and Saturn, the tangible factors of it.

The material world will be an issue whether you are experiencing aspects of the Saturn cycle or those of Saturn to other planets and points. While Jupiter, Uranus, Neptune and Pluto may sometimes be concerned with solely internal developments, Saturn usually denotes concrete, visible results. Even when Saturn is activating the Moon, which represents feelings, there is a need to understand and apply the emotions in practical terms or clearly defined situations.

Saturn to **Sun** aspects advise you that you should consider your ego needs. You might start thinking along these lines because you are not getting the recognition you want, or because you need greater achievement for self-satisfaction. Since one of key words for Saturn is clarification, these aspects can provide a time to define goals, to organize your methodology and to slowly move in the desired direction.

I have a client in the music business who made use of this approach. The young man wants to be a superstar, and his natal chart indicates that he has a tendency to be in a hurry. He was fortunate enough at the age of 21 to be a self-supporting, full-time musician. Everything seemed to be moving in his favor when transiting Saturn conjoined his natal Sun, and his career seemed to come to a halt. His words were, "I'm not getting anywhere in my job. And I don't have the kind of respect I want." I reminded him of where he was compared to others of this age (then 23), and suggested that perhaps because he had moved ahead so quickly, he should take stock of his accomplishments and determine where he needed improvement. The symbolism of Saturn to his Sun suggested that he had to be thorough in order to attain ego gratification. He admitted that there was much more he could learn about the basics of his work, and decided to consider this an apprenticeship period. With this decision he became more patient and worked hard at diligently furthering his training. Several months later he called to tell me that life was treating him much better. He had evidently done his homework well because he was then being allowed to produce and direct as well as perform. I asked if he was getting more respect. He thought for a moment and said, "Yeah, that too."

Saturn to **Moon** may not, at first, be as overtly evident as Saturn to the Sun. You might find that emotions are difficult to express. This may be self-imposed, or you might think that others are forcing you to hold back or are just not interested in your feelings. But, in any case, dealing with them only internally is not enough. There is a need to bring them out into the world. If these aspects initially seem to be experienced as depression and self-pity, my advice is usually to wallow just a little bit. Think of everything you can that does or could depress you, because this is one way to get in touch with what is really troubling you. Then you should set about correcting whatever is wrong. Another possibility with transiting Saturn to natal Moon is duty (Saturn) to mother (Moon), so along with clarifying your own emotions you might want to do something for your mother, or clear up any misunderstandings with her, if applicable. This may seem to be a trivial interpretation, but I cannot omit it because often clients have informed me that during these periods they have been able to alleviate difficulties with their mothers.

When transiting Saturn is aspecting your **Mercury,** you might feel restricted in your communications. You may not be able to say what you are thinking. Or you could complain that no one is listening to you. Or, self-criticism is inhibiting your speaking or writing. Fear of failure leads to blocks in communication. Plan what you want to say and rehearse if there are important points you want to make. This, of course, takes effort, but you will find that during these periods thoughts can be easily clarified, and putting ideas into writing can provide a visual aid in the process. I find that these aspects almost always accompany my urge to write something substantial. In fact, as I am now writing, transiting Saturn is making a station in square to my natal Mercury.

Transiting Saturn to **Venus** is often described as difficulty in expressing affection or limitations in love relations. It can even mean the end of a love relationship. But security and stability are also ascribed to Saturn. You can strive for this, rather than for misery. A student who had a rocky relationship with her boyfriend came to class one night highly distraught. "Wouldn't you know that the time we're supposed to take our vacation together is when Saturn is opposing my Venus. Maybe I should stay home." Someone in the class asked her how she would feel if she stayed home. "Miserable," she replied. "That's one way of fulfilling the expectations of the aspect," said the other student, and continued, "And he might find someone else so that your relationship would end. And that's even a stronger manifestation of the aspect." Those words of wisdom helped her to decide to take the trip, and it turned out differently from the way she had imagined it would. She was determined to use the positive energies represented by Saturn, and while they were away together, she intentionally brought up issues that were disturbing her about the relationship. She said the neutral surroundings made discussion easy, and when she returned she felt much more secure in the relationship. When Saturn to Venus aspects are present at time of marriage, they are often indicative of an enduring relationship.

Saturn to **Mars** aspects should be used to take deliberate action. If you feel barred from moving in the direction you wish to take, reevaluate your situation. You might discover that your choice was not the best or that a few rocks have to be removed before you can continue on the path. Saturn may seem to slow your progress, but in the long run it can offer the most direct route.

Saturn to **Jupiter** aspects might seem to limit your development but should be viewed as times of consolidation. These periods can be used to understand and practically apply newly acquired material. They also symbolize the need to avoid spreading yourself too thin. If you feel frustrated in your attempts to take on new activities, it may be that you are already doing enough or too much, or perhaps you should just consider the value of your aspiration.

With transiting Saturn to **Uranus** you could find that your freedom and independence are being interfered with. Duties and responsibilities or feelings of guilt might make you think you are boxed in. You could just complain about never being able to do what you want to do, and then you probably **will** feel overburdened and frustrated. If, however, you regard these periods as times to determine your responsibilities and find expedient ways to fulfill obligations, you will ultimately feel freer. These aspects can also indicate propitious times to channel creativity and originality.

When transiting Saturn aspects natal **Neptune**, blind faith is not easily accepted. You may want proof to substantiate your spiritual notions; and giant flaws could appear in your belief system. You could be disappointed in the behavior of others, think that people are trying to take advantage of you or question your own altruism. Selecting a project to help some worthy person or cause could allay potential feelings of guilt and resentment because you are being of assistance and your obligations are self-imposed. Any practical expression of the non-material side of your life can be satisfying. For instance, you might join an organized religion that seems to support your views to feel more secure and also to better understand them. Since Neptune can also find its outlet in the arts, you might decide to take dancing or music lessons during one of these periods.

Saturn to **Pluto** aspects can be associated with issues of power. You might feel as though your personal power is being challenged—perhaps someone is trying to control you. You could also seek or attain positions of authority during these periods, but concerted effort may be a part of the process. The more organized you are, however, the more likely you are to succeed. Another element in this combination is getting to the bottom of things. If you try to ignore feelings of dissatisfaction, unpleasant situations are bound to surface. You should, instead, face and clarify problems (Saturn). Then you can eliminate obstacles in order to transform (Pluto) unsatisfactory conditions into more acceptable ones.

As Saturn aspects your **nodes**, you could feel overburdened with responsibility for others, and you might want to end associations with individuals who are sapping your energies. Feelings of potential loss might keep you from taking such action. But if you seek out those on whom you can depend, and solidify **these** relationships, the less desirable ones can more easily be terminated.

If Saturn is aspecting your **Part of Fortune**, you might feel that the road to fulfillment is blocked—that what you want to accomplish is beyond your reach. Your sense of wholeness might be fragmented. On the other hand, if you examine the fragments and understand what assets you have to build on, you can crystallize goals and direction.

Saturn to the **MC** transits are times during which you should be honest and project your best self out into the world. Previous hard work will be rewarded, but dishonesty will be discovered and lack of thoroughness could lead to failure. This is why both promotions and loss of jobs fall under this category of transits. You might find that there is much to be done, but by organizing tasks and finishing one task at a time, you can achieve a great deal.

Saturn to the **Ascendant** aspects are similar to Saturn-MC in that you are being advised to do what is right; but here you are dealing with more personal obligations and direction—taking stock of who you are as an individual and not necessarily your worldly accomplishments. If you are not happy with the personal impression you make, you could just allow yourself to be depressed, or you might use the discipline suggested by Saturn to diet or in some other way improve yourself.

As Saturn moves through the houses, you are afforded the opportunity to organize goals in terms of the area represented. Since transiting Saturn is in a house for about 2-1/2 years there is enough time to work hard for desired goals and to reap the rewards of your labors. A number of years ago, as Saturn was conjoining my Ascendant, I read a description of the Saturn cycle throughout the houses. It was stated in this book as that transiting Saturn enters the first house, one should retreat from the world and prepare to reenter it 14 years later when Saturn crosses the seventh-house cusp. This was quite depressing (a sound negative expression of Saturn conjunct Ascendant) for me, a Leo, since Leos seldom want to crawl into a cave. Subsequently, I discovered that this was not necessary. I found that if I treated each house as a separate unit, improving and solidifying the area as Saturn moved through it, the rest of my life could proceed as usual. Even working on self-improvements (the first house) can be done while you are interacting with the outer world. In fact, it offers immediate feedback, which cannot be found in isolation.

# Alchemy

I have stressed the work element of Saturn, and Saturnian periods are often accompanied by feelings of being overworked. There can seem to be so much to do that you do not know where to begin. So there may be the temptation to do nothing and feel unhappy or project blame on others for your lack of accomplishment. Alchemically this can be easily handled by making lists of tasks that need to be tackled. This type of visualization helps to prioritize. You can then clearly see the jobs that must be done. You can even number them in order of importance. Then concentrate on the first task, and when it is completed cross it off your list. This gives you the incentive to attack the next, and with the elimination of each chore comes a greater and greater sense of satisfaction. Lists can also assist in making decisions, and there always seem to be some decisions that must be finalized during Saturn aspects. Simply write down the pros in one column and the cons in another. You can then weigh one column against the other. First you can compare the length of each list. If you have ten items in the pro column and only four in the con column, this could draw you toward a yes answer. Then evaluate the content of each list and the decision will come so easily that you will wonder why you ever thought it was a problem.

An example of the success of this technique was demonstrated by a young woman who natally has a six-planet stellium in Leo all square Saturn in Scorpio. At the time of her first consultation she had transiting Saturn conjoining her stellium and squaring her Saturn. She was a senior in college and was so overwhelmed by the amount of work she had to do that she doubted that she could make it through the year. When I suggested that she make lists, she thought the idea was a little strange, but she was willing to try anything. She was surprised that in a short period of time everything fell into place. Since this transit activated seven of her ten natal planets and lasted a number of months, one set of lists was not enough. But each time she began to feel depressed or overburdened, she again made lists, and each time it made her feel better and started her moving again. The rewards of that year were the highest grades she had ever received in school, a bachelor's degree, a husband, and a job in which she could utilize her college major. About a year later she told me that that time had been the best of her life. The difficulties she had faced and overcome were no longer present and had faded from her consciousness. Her concrete accomplishments remained and are what she still associates with that period.

Another alchemical application of Saturn literally is to put things in order. Saturn does not function well in chaos. If you clear your work space and find places to store extraneous items, you will find, for some inexplicable reason, that this provides the incentive to work in other ways.

One final facet of Saturn that alchemical rituals can address is that of responsibility. First, you must accept that Saturn is indicative of responsibility. If you shirk your duty, you will undoubtedly feel guilty or dissatisfied, but still you want to avoid being weighted down with too many obligations. If you try to deny the urges that are present and do nothing, very possibly you will only relate to the negativity that can be associated with Saturn. But if you **voluntarily** take on the responsibilities that you wish to assume, you may avoid being asked to do things you prefer not to do. Or if you **are** asked, you can always say, "I'm already committed to something else." This will make saying no easier and you can do it with a clear conscience.

Although the limiting, confining side of Saturn often comes to mind first, remember that Saturn can also be connected with security, accomplishment and success. Yes, Saturn may seem to be restrictive. You might feel as though you have a millstone around your neck, but with a little effort it also can be a rock-foundation.

## Transiting Saturn in Houses

a. Saturn in the 1st house
b. Saturn in the 2nd house
c. Saturn in the 3rd house
d. Saturn in the 4th house
e. Saturn in the 5th house
f. Saturn in the 6th house
g. Saturn in the 7th house
h. Saturn in the 8th house
i. Saturn in the 9th house
j. Saturn in the 10th housee.
k. Saturn in the 11th house
l. Saturn in the 12th house

1. _____ Get sick, make out a schedule for your daily routine, or tackle a difficult task that needs to be done.

2. _____ Close doors that might broaden you, devote yourself to an organized religion, or carefully plan long trips.

3. _____ Complain about never having money, consider expenditures carefully, or put some of your money in the bank.

4. _____ Brood and keep your ideas to yourself, have serious conversations, or write something you consider important.

5. _____ Stifle your children or your creativity, lay down rules and regulations for your children, or start an artistic project that will bring you rewards.

6. _____ Feel limited in your environment, put your house in order, or take on new responsibilities in the home.

7. _____ Refuse help from others, conscientiously handle other people's money, or find security in your sex life.

8. _____ Feel insecure because you think you are an inferior person, work diligently on your appearance, or examine personal responsibility.

9. \_\_\_\_\_ You might feel professionally frustrated and could even lose your job, or place time and effort into your career, and be honest in your dealings with the world.

10. \_\_\_\_\_ Allow yourself to feel depressed and wallow in self-pity, try to understand hidden motivations, or take on responsibilities in an institution.

11. \_\_\_\_\_ Be unhappy with your partner, commit yourself to a serious relationship, or examine problems that could interfere with your partnership.

12. \_\_\_\_\_ Avoid groups entirely, set limits on your social life, or teach a subject you know well to a group.

# Transiting Saturn in Aspect

The following exercise includes possible uses of transiting Saturn motifs as Saturn aspects planets and points in the natal chart.

a. Saturn-Sun
b. Saturn-Moon
c. Saturn-Mercury
d. Saturn-Venus
e. Saturn-Mars
f. Saturn-Jupiter
g. Saturn-Saturn
h. Saturn-Uranus
i. Saturn-Neptune
j. Saturn-Pluto
k. Saturn-MC
l. Saturn-Ascendant
m. Saturn-Part of Fortune
n. Saturn-nodes

1. \_\_\_\_\_ Feel overwhelmed by all your responsibilities or examine your life structure in terms of goals and direction.

2. \_\_\_\_\_ Avoid people or take on responsibilities in a relationship.

3. \_\_\_\_\_ Wallow in self-pity or carefully examine your emotional needs.

4. \_\_\_\_\_ Decide that personal fulfillment is impossible and do nothing, or determine a direction that will give you personal fulfillment.

5. \_\_\_\_\_ Feel depressed about the way you look or go on a diet.

6. \_\_\_\_\_ Be unhappy because everyone uses you or work hard for ego gratification.

7. _____ Be restricted in expression of affection or put a love relationship in order.

8. _____ Do only your duty and suppress your urge for freedom or allow time for expressing your individuality.

9. _____ Stop working at your job so that your boss will have reason to fire you or work hard on your career.

10. _____ Keep quiet and grumble about no one listening to you or carefully plan out what you have to say.

11. _____ Let energy build up until you explode or get involved in planned physical activity.

12. _____ Ignore dreams, however significant or utilize dreams to reach your goals.

13. _____ Punch someone in the nose, or probe deeply to get to the bottom of problems.

14. _____ Give up any idea of broadening your horizons or take action on plans for growth and development.

## Chapter Six

⛢

# Uranus

A Uranus cycle takes 84.02 years. It is in a sign for about seven years and moves a maximum of 4 minutes a day. Therefore an aspect is usually in orb for about a month.

## Function

Uranus signifies change, independence, creativity and originality. All transiting planets can be connected with some type of change in our lives, but Uranus is the planet most closely associated with those that are sudden and abrupt.

## How You May Feel with Uranus Transits

There is an urgency connected with Uranus. As it forms aspects with your natal planets and points you might feel a need to alter your environment or your attitude. Ideas will enter your mind that you will want to act upon immediately, and you will usually not have the patience for thorough investigation. It shares this quality with Mars, but Mars aspects are in range for shorter periods of time, so the possible carelessness and impatience common to them both are often more evident with Uranus.

Uranian periods can be stimulating but may be unstable. The emphasis is on making things different, but the direction is not always clear. You may change your mind frequently, and sometimes even question your own sanity because you can be so inconsistent. Your mind will seem to be in constant motion, and you will have brilliant thoughts along with some strange ones. You can be extremely creative, or downright bizarre.

You might also feel restless and crave excitement. If you are not actively expending energy, you could become nervous and erratic in your behavior. You want to express your individuality and independence. This can take the form of assertiveness to attain freedom, or of purposeless actions just to be different.

## How Uranus Transits May Be Used

If there are changes that you have been planning to make for some time, you could select Uranus transits to initiate them. Be cautious, however, about closing doors behind you before the path ahead is cleared. Consider these experimental periods and test the water before you dive in head first. Otherwise the consequences may be less than desirable. An example of this was described to me by a client who was born in 1952 with Saturn conjunct Neptune in the seventh house of her natal chart. When transiting Uranus was conjoining this conjunction, she awoke one morning with the impulsive notion that she had to live with her boyfriend in California. She quit her job that day. She then returned home, packed her bags and flew to California. She thought it would be a wonderful surprise for her boyfriend. But when she arrived at his apartment, the surprise was hers. He was living with another woman. It took her months to get her life back in order.

The message of the transit was clear. She had to change (Uranus) a relationship (seventh house) by dissolving (Neptune) the structure (Saturn) and forming a new one. But there were alternatives as to how this could have been done. Had she first taken vacation time to visit him without severing all her other ties, the relationship might have ended in any case, but she could have avoided the complete disruption of her life. It is not necessary to instantaneously and totally change your whole life under Uranus transits. There are always choices. And the better you understand the principles of the planets, the more effectively you can apply their symbolism.

Everyone experiences the hard aspects of the Uranus cycle at approximately the same ages. The first square of transiting Uranus to itself in the natal chart occurs somewhere between 18 and 23. If you were born in August of 1924, you would have had this aspect at age 22. If you were born in August of 1944, Uranus square Uranus would have been exact at age 20. Whenever this aspect formed, there is the desire for change and more independence. According to *The Seasons of a Man's Life* the **Early Adult Transition** which takes place at ages 17-22, "…is a time of profound change in self and world" (p.78). It is further stated that "In the Early Adult Transition, the adolescent-becoming-adult has a special concern for his own independence as he struggles to pull away more completely from parents and from the pre-adult self that is still so strongly tied to them." (p. 144). The Uranian tasks of this period, whatever your circumstances, are to become more independent, and also to explore the possibilities of the adult world— "…to make and test some preliminary choices for adult living" (pp. 56-57).

The next hard aspect of the Uranus cycle (Uranus opposition Uranus) is formed between 39 and 43 – again depending on the particular year of birth. This aspect extends from the end of the **Settling Down** period (33-40) into the **Mid-life Transition** (40-45). The early **Settling Down** period, which contains hard aspects of the Saturn cycle (but not of the Uranus cycle), "is devoted to building a life around initial choices. It is a time for making one's niche in society, defining an enterprise, getting on with the work, 'taking care of business'" (p. 143). These, of course, are all Saturnian concepts, but according to the book, in the late **Settling Down** Period (during the sub-period of **Becoming One's Own Man**) "…a man wants to be more independent, more true to himself and less vulnerable to pressures and blandishments from others.… The wish for independence leads him to do what he alone considers most essential, regardless of consequences…" (p. 144), Obviously Uranus has entered the scene. And then, in the **Mid-life Transition**, the Uranus opposition Uranus task is "…to modify the negative elements of the present structure and to test new choices" (p. 192).

You will note that the psychologists use such terms as "exploring" and "testing" during periods in which Uranus is in the foreground. This corroborates the experimental nature of Uranian periods. Try not to make irreversible decisions during these times. Do your testing with the knowledge that more changes can still be made.

I find that clients experiencing Uranus opposition Uranus frequently discuss change and freedom as issues in their lives during this transit. Statements about it vary from "my children have left home and what am I going to do with all my free time" to "I am unhappy in my job and want to leave it" to "something has to change in my marriage or I want out." You may have been completely content before this time, and suddenly restlessness or dissatisfaction with some segment of your life emerges. Sometimes there is the mention of life being half over and of time having been wasted. Or a client will ask "What is wrong with me that I am no longer happy with my life?" My stock answer is "There is nothing wrong with you, and you have not wasted your time. You have been growing, and situations that were once right for you might need to be changed because of your development. Without previous experience how could you properly judge the direction to take?" This always seems to relax the client.

Some individuals flow with Uranian issues more easily than others. I had a client who was having great difficulty adjusting to what was occurring during her Uranus opposition Uranus. Her goals and direction had always been clear. She wanted to own her own restaurant and she had done this by the age of 35. She had attained local recognition as a gourmet cook and her business was successful. How could she want anything more? But at 40, when transiting Uranus opposed itself in her natal chart, she became bored and unhappy. Her first words when she arrived were: "I think I am going crazy. I have everything I always wanted and suddenly it isn't enough. I have this strong urge to sell my business. But then what would I do? I am so confused. But I have to do something."

I first reassured her that she was not out of her mind, gave her my standard answer about change and development, and then suggested that there were certain steps that could be taken without resorting to drastic action. My advise was not to dwell on whether or not to sell her restaurant at that moment, but rather to concentrate on being creative. Then, perhaps, the decision could more easily be made. This alchemical ritual usually works. In her case, she began to invent new dishes, and as she added these to her menu, she became less bored with her life and her business.

With the second Uranus square Uranus in one's early sixties, and the conjunction at around age 84, similar Uranian issues again become prevalent. In the early sixties, for the working person, retirement can become a focal point. Frequently individuals are thinking about the freedom they are going to have, and can view this idea with excitement or dread. I have a number of clients who actually started new professions—often quite different from their previous ones – during this period.

At the Uranus Return (around age 84) independence can come to the foreground again. At this age health is sometimes a factor. I had a neighbor who at the time of her Uranus return came to my house in tears. Her son had just told her that he wanted to put her in a "home." She lived alone and was totally self-sufficient. Up to this time she had never considered herself old. In fact, one of her favorite pastimes was playing the piano for the senior citizens in order, according to her, to bring a little pleasure into the old people's lives! Her son's concern was that, since she lived alone, no one would be around to help her if she suddenly became ill. All of the neighbors met and decided to take turns checking in on her each day. This relieved her son's mind, and she was able to maintain her independence.

As Uranus aspects the other planets and points in the natal chart, the need for excitement, freedom and change will be exhibited, but the desire for change or independence is focused on the function or segment of character that is represented by the planet or point being activated by transiting Uranus.

With transiting Uranus aspecting natal **Sun** you could become dissatisfied with the ego gratification you are receiving. You may want to "shine" in a new way. You now need to be recognized **as an individual**, so establishing or asserting your own identity can be important. The dutiful, homebound mother might join a woman's lib group; or the usually docile employee might start a revolution at work. I was once told a story of a man who worked on an assembly line in an automobile factory. His job was to put the doors together. He performed this same task day after day and was just a small cog in a very large machine. One afternoon he took a handful of screws and threw them between the inner and outer door parts. He spent the rest of the day doing this with each door. I'm sure it made him feel as though he had placed his personal mark upon the world until customers began to complain and it was traced back to him. He had expressed his individuality, but ultimately he was fired. The point of the tale is that there are bad choices as well as acceptable ones that can satisfy your need for personal recognition, and you might want to consider the consequences before you take action on your particular decision.

Transiting Uranus activating your **Moon** generally indicates fluctuation of emotions. Even the most stable individuals are subject to sudden mood swings during these aspects. This can be particularly unnerving to those who have little water in their horoscopes. One such client found an acceptable release, which I have added to my alchemical list, and that was to watch sad movies. This provided a reason for the tears to flow and he found that relieving himself of these feelings helped him to more objectively deal with problems he was having with his wife at the time.

During Uranus to **Mercury** aspects you will probably notice that thought processes are accelerated. You could find yourself communicating spontaneously without considering what you are saying—another possible "foot-in-mouth" combination. You are also apt to change your mind frequently. Ideas will often come in flashes, and there is not always the time or patience to develop them. During such periods you should jot down thoughts or put them into a tape recorder for later use. I have a client who is a free-lance writer. When Uranus was squaring her **Mercury**, she complained that she was sleeping fitfully because she was periodically waking up with superb ideas for articles or stories. She was too tired to write them down immediately and each morning she would awaken exhausted with the thoughts having evaporated from her memory. After her consultation she began to keep a tape recorder next to her bed. This required that she merely push a button. In the morning she would listen to the tape and sort out the material that she wanted to use. She also found that her sleep was less often interrupted and that she was not as exhausted and nervous as she had been previously.

With Uranus to **Venus** you can be creative in artistic endeavors. If you have marked talent in this direction, you could produce noteworthy, original works during these periods. If you have never tested your artistic ability, you might choose such times to investigate one of the arts. You should not concentrate on methodology or strict discipline but rather let your energy flow freely. Also, do this only when you feel inspired. If you try to push when you are not in the mood, you will probably not accomplish anything. Another facet of this combination could be love or expression of affection. You could crave excitement in a love relationship. Since a Uranus-Venus combination can be indicative of "love at first sight," if you are unattached, you might expose yourself to situations in which you could meet someone of interest during these aspects. Go to parties or take a cruise. If you **are** romantically involved, you could become bored with the relationship and want it to change in some way. Before you take drastic action, create a little excitement. Plan a romantic evening or try a new and potentially creative activity with your partner. This may be all that is needed to bring back the spark. Or if you have been inhibited in a love relationship you could find that this is a time to become freer or more open.

Uranus to **Mars** aspects, like Mars to Uranus, can be times of impatience and possible carelessness. You might as well plan energetic activities during these periods because it will probably be difficult to sit still. Powers of concentration are limited, so keep tasks short. Alternate physical and mental action, and if your mind begins to wander, shift your mode of operation. Even the most timid individuals can be sporadically self-assertive. If you fall into this category, you could select such times to take the initiative and express personal needs that have been sublimated.

During Uranus to **Jupiter** aspects you should look for new ways to grow and develop. You want quick results, however, so this is not the time to institute long-range plans unless it is otherwise indicated or, unless periodic progress can be measured. An astrologer friend took a cross-country lecture tour while transiting Uranus aspected her Jupiter. She moved quickly from city to city and had a different setting for each lecture, In this way, she could evaluate each talk as an isolated unit and maintain an air of excitement through different audiences and topics. But there was still an overall accomplishment that was much broader – she became more nationally known and made contacts for future activities.

While Uranus is aspecting your **Saturn**, you will probably feel the urge to rid yourself of responsibilities, or at least change the ones you have. You can also become aware of, and dissatisfied with, your own limitations. This is a good time to break patterns, but you should proceed cautiously. Although, Uranus may seem to demand immediate change, you do not have to make irreversible alterations abruptly. A new pattern should have a purpose. Natal Saturn can assist you in determining this. While transiting Uranus is attempting to pull you in a totally new direction, natal Saturn tends to resist change. So you can consider this combination as a "tug of war" or a "balancing act." If you consciously make small changes, you could avoid extremes. Since the emphasis is on some segment of your life structure, you might have the urge to quit your job. Before you take such action, however, examine what is wrong with present conditions and try to correct problems. If you leave your current position without any forethought, you could find yourself without means of support or in an equally unsatisfying job because you acted too hastily. First investigate what is really disturbing you about the job. When the root of your dissatisfaction is clear, you might discover that only minor changes are needed. If, however, the flaws of your present situation are too great to be reconciled, you can leave with the knowledge that your decision is right and with the awareness of what you want to avoid in your next position. This approach is applicable to any segment of your life.

Uranus to **Neptune** aspects are much less tangible that those of Uranus to Saturn. Neptune is not associated with the resistance or grounding that Saturn signifies. The urge for change associated with Uranus is coupled with the nonmaterial, spiritual quality ascribed to Neptune. Therefore, you may be inconsistent or feel confused about what needs to be done. You could have a sudden urge to become involved in religion or philosophy, and just as quickly change your mind. You could have psychic flashes, but these will be sporadic. Another possibility is that you could want to free yourself (Uranus) from some form of victimization (Neptune). Again there is not a great deal of consistency, but under these aspects you could periodically rebel against those whom you feel are trying to take advantage of you. This combination, however, can be activated in a totally different way. Since Neptune is the higher echelon of Venus, and Uranus represents, among other things, creativity, you could use these aspects for creative projects.

You could select the combination of Uranus and **Pluto** as a time to start a revolution since Uranus represents change and Pluto transformation. With Uranus as the transiting planet the uprisings might occur suddenly and end just as abruptly. On the personal level the appropriate application of these aspects is periodically to assert your own power, especially if you are feeling threatened by a strong individual. A mild-mannered client with Pluto in the ninth house natally, found, much to her surprise, that when transiting Uranus squared her Pluto, she began to challenge her domineering sister-in-law. Her initial reaction, as often occurs with Uranus aspects, was "What's wrong with me?" Before this time she was not consciously aware that her sister-in-law disturbed her at all, but during this period she was often contradicting this dogmatic woman even when she really did not disagree with her. I suggested that perhaps these outbursts were indicative of an unconscious problem with the relationship from her perspective, and that she should make use of the deeply probing function of Pluto to analyze the situation. In this way she might face and change her manner of interaction with her sister-in-law. Ultimately she determined that she resented the control his sister had over her husband. This helped her to combat her own argumentativeness (with which she felt uncomfortable) and direct her energies more effectively. These aspects, however, are not always connected with relationships. Without external stimuli you may have the urge to get in touch with your own power, or feel the need to bring psychological problems to the surface in order to eliminate them.

With Uranus activating the **nodes**, on the other hand, relationships are always relevant. You might find yourself dissatisfied with the manner in which you interact, or with those with whom you come into contact. New associations could be formed suddenly or old alliances ended just as quickly. Unusual or creative people may enter your life. Clients have frequently told me that with these aspects individuals from the past often unexpectedly materialize. Since you attract original or unusual individuals during these periods, you might choose to seek not those who are creative, and thereby, hopefully, avoid those who are less desirable. In an existing relationship, if you are bored, or feel that your individuality is being threatened, try some new activity with this person, or assert your independence.

As Uranus aspects your **Part of Fortune**, you could feel fragmented, or as though something is missing from your life. Wholeness and fulfillment **will** be crucial issues, but you might often change your mind as to what these entail. Allow yourself some experimentation during Uranus to Part of Fortune aspects. With the idea of fulfillment in the back of your mind, seize any opportunity that might satisfy this need, or make your own opportunities. There could be some false starts, but making an effort is certainly preferable to the alternative of restlessness and dissatisfaction with oneself. When transiting Uranus conjoined my natal Part of Fortune, I began to study astrology. Occasionally, in those early days, I sometimes wondered why I was doing it. But, by the time Uranus moved out of range of my Part of Fortune, I was totally absorbed in the subject. And it has definitely brought me a sense of wholeness and purpose.

A Uranus to **MC** aspect can be a propitious time to make a career change if you have been contemplating it prior to this period. You could send out resumes, investigate possible openings and then, when the aspect is in orb, make your move. If you are considering a totally different type of career, you might look deeply into the new field, and at the time of the aspect either change occupations or change your mind. If, however, you have been completely satisfied with your present position, and, as the aspect occurs, suddenly have the urge to leave it behind, be a little cautious. First try to make changes within the job. Express your individuality, become involved in creative projects or reorganize your work schedule to relieve potential monotony. This may be all that is necessary. Since the MC also represents interaction with the world-at-large, you might choose to express your individuality and need for freedom through some other role you perform.

During transiting Uranus to **Ascendant** aspects you might experience a general feeling of restlessness and strong desire for more freedom regardless of your life circumstances. A mundane example of this, if you use the Gemini-rising US chart, is that the Civil War was in progress when transiting Uranus conjoined the US Ascendant. An issue of this war, of course, was **freeing** the slaves. I do not wish to start a controversy here – so if you do not adhere to the Gemini-rising chart, let me say that you could associate the war with the Uranus return. I submit, however, that the mood of the country during the Civil War is applicable to the individual when transiting Uranus is activating the natal Ascendant—a time of unrest and strong opinions on matters of freedom. With these aspects the need for change may emerge in different areas and may shift from one to another. Perhaps, then it is best to start with small changes. The Alchemy section offers some suggestions.

As Uranus transits each house, you will have approximately seven years in which to make changes in that area, or to modify your approach to that segment of your life. You will probably not be exposed to seven years of constant turmoil, but expecting the unexpected will help to prepare you for sudden turns of events which might occur occasionally in terms of that house. Better yet, you might periodically consider what you would like to do differently, or how you could be more creative or individualistic in that area. Then consciously make an effort to move in that direction. By channeling the Uranian energies into activities and issues beneficial to you, you could not only find the area more satisfying after the transit, but you might also avoid some unpleasant surprises. There will undoubtedly be some changes in that house or your method of operating there, so you might as well take advantage of the opportunity to set the new scene.

## Alchemy

To make the best use of Uranian themes alchemically, you first should develop the appropriate attitude. This, of course, is true of any planet. You must accept the reality that you cannot, for instance, deal with Uranus as if it were Saturn, Uranus aspects cannot be used to set clearly defined parameters or solidify your position. If you insist upon trying to do this, you will probably feel unsettled and nervous, and be frustrated in your attempts. Instead, you should elicit originality and spontaneity within yourself, and then you can produce very positive results. Since all the planets are in continual motion, you can still structure the area represented by the house being transited by Saturn, or the facet of character symbolized by the planets that Saturn is aspecting, but wherever transiting Uranus is—"hang loose." So much time can be wasted trying to determine what is wrong with you under Uranus aspects that golden opportunities might be missed. One way to combat this tendency is to examine transiting Saturn's role, and to stabilize **that** segment of your life. Then you can permit yourself to express individuality in the Uranian sector.

It is wise, however, to avoid total domination by Uranus. If you pursue each Uranian whim for drastic change, you might find yourself in untenable situations as did the woman, mentioned above, who uprooted herself without thought of consequences and moved to California. So you should develop some alchemical rituals that will temporarily alleviate the restlessness and urgency connected with Uranus while you carefully consider more meaningful changes. Some of these have already been discussed. Involve yourself in creative projects, but do this without setting firm deadlines. The creativity will come in spurts. Use it when it is there but do not try to force it when it is in abeyance. If you persist in pressuring yourself when you are uninspired, you will produce very little of value.

Seek acceptable ways to express your individuality and satisfy the need for excitement. If you begin to feel stifled because you do not have the freedom that you want, and are not sure that you should remove yourself from a given situation which seems to be causing these feelings, simply take a little time off and do something for yourself. One client went on a series of shopping sprees by herself, and only for herself, during such a transit. Then she comfortably, and without resentment, went home and handled her responsibilities. She was born with a Saturn-Uranus conjunction, and the freedom-responsibility dilemma is frequently an issue for her. She is also in a position in which she can afford these shopping expeditions. Other alternatives are: becoming involved in a cause, taking on new activities, or in some other way creating a little excitement without disrupting your present life circumstances. Then you can make serious changes when the dust settles, and be more certain that they are right for you.

You will probably notice that something in your life is different after Uranus transits. The changes may be visible or internal, but they should be evident at least to you. You will not be unrecognizable to those who know you, nor will you become someone who belies your natal chart. And, if you have been trying consciously to direct the aspects, the end result will undoubtedly be an improvement over the old self.

## Transiting Uranus in Houses

a. Uranus in the 1st house
b. Uranus in the 2nd house
c. Uranus in the 3rd house
d. Uranus in the 4th house
e. Uranus in the 5th house
f. Uranus in the 6th house
g. Uranus in the 7th house
h. Uranus in the 8th house
i. Uranus in the 9th house
j. Uranus in the 10th house
k. Uranus in the 11th house
l. Uranus in the 12th house

1. \_\_\_\_\_ Start a revolution in a foreign county, take a long trip on impulse, or look into an unusual religion or philosophy.

2. \_\_\_\_\_ Look for unreliable lovers, give your children some freedom, or take up a hobby that can be creative.

3. \_\_\_\_\_ Say whatever comes into your head, try to communicate in a new way, or express unusual ideas and keep an open mind.

4. \_\_\_\_\_ Change careers frequently, or try to become more creative and independent in your profession.

5. \_\_\_\_\_ Associate with weird and unreliable people, join a group activity that involves creativity, or leave a group that has been restricting.

6. \_\_\_\_\_ Run away from home, make changes in the décor of your house, or move.

7. \_\_\_\_\_ Act strangely, assert your personal independence, or change your appearance in some way.

8. \_\_\_\_\_ Concentrate more on daily excitement than accomplishment, or change or add a new and different activity to your routine.

9. \_\_\_\_\_ Change partners, often, express your individuality in your partnership, or be more spontaneous with your partner.

10. \_\_\_\_\_ Spend money carelessly or spend your money spontaneously on something that will make you feel freer.

11. \_\_\_\_\_ Repress your urge for freedom, study a mystical subject, or free yourself from some kind of confinement.

12. \_\_\_\_\_ Handle other people's money recklessly, rid your self of financial obligations, or share creative resources.

## Transiting Uranus in Aspect

The following exercise includes possible uses of transiting Uranus motifs as Uranus aspects planets and points in the natal chart.

a. Uranus-Sun
b. Uranus-Moon
c. Uranus-Mercury
d. Uranus-Venus
e. Uranus-Mars
f. Uranus-Jupiter
g. Uranus-Saturn
h. Uranus-Uranus
i. Uranus-Neptune
j. Uranus-Pluto
k. Uranus-MC
l. Uranus-Ascendant
m. Uranus-Part of Fortune
n. Uranus-nodes

1. \_\_\_\_\_ Say anything to be different or express creative ideas.

2. \_\_\_\_\_ Quit your job so that you don't have to worry about working, or try to attain more independence in your career

3. _____ Rid yourself of all responsibilities or just those which are unnecessary.

4. _____ Dress weirdly or express your need for personal independence.

5. _____ Totally kick over the traces or find new ways to satisfy your need for freedom.

6. _____ Associate with those who are unreliable or seek out relationships with unusual or creative people.

7. _____ Take action without forethought or initiate unusual and creative projects.

8. _____ Decide you are crazy and have yourself committed, or take advantage of insights that come from deep probing although they may be sporadic.

9. _____ Decide it is better to be free than to have fulfillment or find a new way to attain fulfillment.

10. _____ Sit back and wait for a windfall, or try a new and different way to grow and develop.

11. _____ Have strange dreams or look for a new religious experience.

12. _____ Let your moods go up and down or express emotional needs

13. _____ Consider your personal needs more important than anyone else's or use originality in gratifying your ego.

14. _____ Fall in love frequently or express your individuality in a love relationship.

# Chapter Seven

## ♆

# Neptune

The Neptune cycle is 164.79 years in duration. It is in a sign for about 14 years and moves a maximum of 2' a day. Its aspects are therefore in an orb of 1 degree for approximately 60 days. Actually, however, Neptunian periods are much longer lasting. Because of Neptune's slow movement, its symbolism may linger for a long as one to two years. Even when it goes retrograde and moves several degrees from a planet or point it has activated, it does not disappear. As Neptune makes its last pass over the planet or point, the Neptune motifs will slowly begin to wane. This is true also of Saturn, Uranus and Pluto. The four outer most planets seem to flavor periods for much longer than their 1 degree orb would indicate.

## Function

Neptune advises us to become more highly evolved. We should tap into the collective unconscious and be more at one with the universe. It operates on planes that are below and above our physical existence. It forces us to look beneath the superficialities of our lives in the hope that we will reach beyond the material to the spiritual.

# How You Might Feel with Neptune Transits

Since we are dealing with the ethereal under Neptune, it is not strange that the world may take on a mystical or vague aura during Neptune transits. Situations that have previously seemed secure and crystal clear can begin to dissolve. While Saturn defines circumstances in terms of black and white, Neptune seems to cast a gray hue over the total environment. This can lead to a degree of confusion (sometimes referred to as "the Neptunian fog"), especially if you are trying too hard to "make sense" out of what is occurring. In fact, you may literally feel as if you have cobwebs in your head if you attempt this for extended periods of time.

Your reality may contain extremes that shift rapidly or coexist when Neptune is in the picture. For instance, at one moment you might experience a sense of serenity and in the next find that you must deal with pettiness and malice coming from others or from within yourself. Or you may feel peaceful and at one with the cosmos in one segment of your life and simultaneously be plagued by imperfection and dissatisfaction in other areas. As you are searching for the idea, you could question the motivation of those with whom you associate and/or be concerned with deception and victimization.

Since themes of perfection and disillusionment can often emerge simultaneously or shift suddenly during Neptune transits, perfect plans may be created, ideal relationships formed and long-term goals attained. But on the other hand, flaws in existing schemes, associations and aspirations can become glaring. Previous accomplishments may, in retrospect, seem disappointing. This is especially true of material gains. During Neptune periods individuals will frequently ask, "Why did I waste my time working so hard for this goal which now seems meaningless?" The answer is that no achievement is a waste of time. Often, it is necessary to acquire tangible rewards before you can reach the heights that Neptune suggests. You must be familiar with the earth beneath your feet before you can become attuned to the universe.

Although rational thinking may seem impaired during Neptune transits, intuition flows freely. You could be inspired under Neptune and you might also have a strong desire to investigate religion, philosophy or the arts. If you pursue spiritual or artistic goals, you could sometimes feel that you are a vehicle through which ideas and energies not connected with your personal self are channeled.

A desire to escape from your humdrum existence or from situations in which you feel oppressed or victimized might also accompany Neptune transits. Even individuals who are usually realists may create illusions, daydreaming about where they would rather be or what they would rather be doing. As someone experiencing Neptune square Neptune told me recently, "I'd run away in a minute if I knew where to go!" Alchemically daydreams can be a form of temporary relief, and positive visualization may help to alleviate difficulties.

With visualization you picture a situation the way you would like it to be. You want to improve your position but not at the expense of anyone else. You might practice several scenarios until one feels right. Then repeat this visualization at least once a day until the matters with which you are concerned begin to improve. The advantage of this technique is that it is private and personal. If it fails, no one knows but you. And it very often does succeed.

## How Neptune Transits May Be Used

It may be evident by now that the best possible application of Neptune transits is to develop spiritually and aesthetically. Harmonious attunement with the universe should be cultivated as you attempt to activate the energies ascribed to Neptune. Follow spiritual and artistic inclinations. Create dreams, search for beauty and truth, but, as with Uranus, try not to make unalterable decisions on the mundane level.

You can use Neptune transits to become more whole internally. Merging and blending polarities is particularly essential in the **Mid-life Transition** (age 40-45) during which everyone experiences transiting Neptune square natal Neptune (ages 39-43). Polarities mentioned in *Seasons of a Man's Life* are feminine/masculine, destruction/creation and attachment/separateness. Usually one polarity is more highly developed than the other, and during the **Mid-life Transition** there is the need or desire to get in touch with the neglected polarity. The cosmos is all-inclusive, and Neptune transits advise us to reflect this universality. Therefore, at these times we are supposed to become more complete within ourselves.

According to Levinson, men and women in our society are programmed in their early years to believe that certain behavior is appropriate for their gender. Men are supposed to be physically strong and intellectual, while women are assumed to be physically weak and emotional. Of course, all of these opposite tendencies are within every human being. As the more dormant polarity is developed, the character becomes more well-rounded and one is better able to relate to the opposite sex because there is greater understanding.

To balance the destruction/creation polarity, you become cognizant of, and dissatisfied with, the imperfections in your life structure. You might dwell on past acts in which you felt you were harmful to yourself or others, or you might consider this tendency in others. The rationale is that destructiveness interrupts creative energies. As you dissolve these undesirable qualities, you become more productive.

The attachment/detachment polarity deals with our relationship to society. "During the Mid-life Transition, a man needs to reduce his heavy involvement in the external world. To do the work of reappraisal and de-illusionment…He is less dependent upon tribal rewards, more questioning of tribal values, more able to look at life from a universalistic perspective" (pp 241-42). (Although the study on which this book is based included only men, the findings are applicable to women as well.) This may be why previous material accomplishments seem disappointing or meaningless during this period. Attaining "tribal" approval is no longer as important as it was previously.

Another Neptunian theme prominent in the **Mid-life Transition** is the dream. The dream is an integral part of adult development and is therefore discussed throughout the book. But the psychologists include conscious and tangible aspirations (Saturnian implications) under this heading, as well as the illusionary and less rational (Neptunian) ones. Goals such as material success in career, a large and beautiful house, etc., are lumped together with a "noble quest," making the world a better place to live in, finding inner peace, etc. Prior to the **Mid-life Transition** the materially oriented facet of dreams is emphasized. In the **Mid-life Transition** (period in which Neptune square Neptune enters the scene), however, it is the illusionary side of the dream that is heavily stressed. Flaws and failures connected with dreams and aspirations surface at this time, and facing these can help to modify the life structure. Along with recognition of imperfections we become aware of our own mortality, so there can be concern not only with our weaknesses but also with the possibility of death.

The principles that are evident during **Neptune square Neptune** are also applicable as transiting Neptune activates other planets and points, but with emphasis on the facet of character represented by the planet or point being aspected. You will probably be faced with extremes and polarities, and you should concentrate on dissolving or rising above the material in terms of the natal planet or point involved.

As Neptune aspects the natal **Sun**, you are being advised to be less self-centered. This combination is frequently found in the horoscopes of religious figures and spiritual leaders. During these aspects you could be very charismatic. You might find that people are drawn to you for some inexplicable reason. However, it is also possible that your ego will be squelched. Most likely it will be some of each, no matter what the aspect is. You could find yourself swaying between thinking you must be marvelous to wondering if you have any worth at all. With Neptune to Sun transits you should search for ways to be altruistic. Consider these periods as times to give rather than to receive. If you are trying to be selfless, a stronger sense of personal identity could result.

The extremes that may be evident during Neptune to **Moon** aspects are emotional peace and confusion of feelings. You might not always understand your emotions at these times. Incomprehensible feelings of resentment could emerge. Irrational grudges could surface. Trying to understand or explain the emotions rationally will probably be impossible. Therefore, it is better to actively follow some spiritual pursuit, rather than attempt to face negative feelings directly. This may seem to be a Neptunian escape mechanism in which reality is ignored, distorted or sublimated; but actually it is not. By finding ways to express love and peace, feelings of anger and hatred are more easily displaced than by acknowledging them.

An example of this approach came to me via a client who, when Neptune conjoined her Moon, came for an initial consultation. When I discussed this aspect, I mentioned the possibility of feeling emotionally drained or taken advantage of; and that seemed to "push a button." With tremendous anger she began to tell me about her teenage daughter. She said, "I know a mother shouldn't hate her own child, but I can't help myself. This girl is so selfish. She uses everybody, but especially me." She was so upset that I knew it would be useless to say that we should look at the situation objectively. Instead, I suggested that she make use of another side of this aspect, and that was to get involved in some spiritual activity. I explained that this might not change her relationship with her daughter, but it might provide some peace of mind generally. Several months later, when she returned, she seemed such more relaxed than in our previous encounter. I asked if the situation with her daughter had improved. Her answer was "No. But I realize that in the total scheme of things it is very unimportant." Between readings she had joined and become active in a metaphysical group, and definitely had a new outlook on life.

With Neptune to **Mercury** aspects you could notice that communications flow magically, without effort; or you could feel that no one understands what you are trying to say. It is indeed possible that your thoughts **are** confused, especially if you are trying to be logical. Although down-to-earth conversing and practical thinking are not strong during these periods, imagination is. Therefore this combination is excellent for advertising, writing fiction, and spreading propaganda. Since it is difficult to perceive all the practical implications of one's situation, I often suggest to individuals experiencing Neptune to Mercury aspects that they begin studying the *I Ching*. Insights and practical answers that need to be found during these potentially unclear periods may be more easily attained through this book. Prayer and meditation are other beneficial applications of the Neptune-Mercury combination.

In a natal chart a Neptune-**Venus** connection can be indicative of artistic ability. Therefore a manifestation of transiting Neptune activating natal Venus may be the urge to do something artistic. That urge should be followed even if the horoscope does not indicate such talents, because being occupied in this manner may forestall some of the negative possibilities of these aspects. Another positive potential of transiting Neptune to Venus is finding the ideal love-relationship. Since the objective world may be distorted or blurred during these times, you should accept and enjoy the glamour or seeming perfection connected with such a relationship without necessarily making a long-term commitment. It may turn out very well, but you should also recognize the possibility that the "rosy glow" might not last forever. The negative side of these aspects could be feelings of victimization in a love relationship. These may be real or imagined, so rather than dwell on this, you should seek positive ways to express the Neptune to Venus energies.

One client, as transiting Neptune opposed her natal Venus, began to suspect that her husband was having an affair. While she was trying to decide what to do about her suspicions, a unique opportunity was offered to her. The minister of her church asked her if she would teach a Bible class for women. The preoccupation with her marital situation made her want to reject this added responsibility, but when we discussed the aptness of the symbolism (Neptune to Venus could just as easily be teaching the Bible [Neptune] to women [Venus], as it could be victimization in love), she decided to give it a try. It turned out to be a very rewarding experience. She kept finding references in the Bible that not only proved to be good material for her course, but also were applicable to her personal life. Her teaching did not totally eradicate her feelings of uneasiness about her husband, but the knowledge that she acquired ultimately helped her to deal more calmly and effectively with the relationship.

During Neptune to **Mars** aspects you can soften the manner in which you take initiative or assert yourself. Even the most aggressive individuals may seem gentler during these periods. These aspects can also be excellent times for expressing grace and beauty in action through such activities as skiing, skating, diving, and dance (all of which involve beauty as well as physical exertion). On the other hand, these aspects might coincide with dissipation of physical energy. You could find yourself tiring easily. Or you might be involved in situations in which it is difficult to take direct action. An overprotective mother found herself in such a position with the authorities and she wanted to rectify the situation, but there was nothing she could do. Ultimately her son handled it himself, and I believe, as a result, he became stronger. So a positive Neptune to Mars message can be that aggressive behavior is not always necessary and that sometimes it is best under such circumstances to let someone else take the initiative.

Transiting Neptune aspecting natal **Jupiter** may be used to acquire knowledge in the realm of religion or philosophy. As with Jupiter to Neptune, however, Neptune or Jupiter is not the best combination to cope practically with objective reality. So, here too, you should be wary of drugs and alcohol, which may distort your perception of the world around you. You should enrich your spiritual life as an adjunct to your daily existence, but not as a substitute for it.

When Neptune activates **Saturn**, you **are** being advised to examine your material structure so that you can rise above, dissolve or perfect matters that are impeding your progress. These aspects may emerge as depression and dissatisfaction because of awareness of flaws within yourself or your environment. Or you may simply have the desire to enhance your physical world through spiritual development without inner pressure. Neptune to Saturn aspects can make you realize that material accomplishments are not always enough.

Saturn represents security as well as physical structure and where it is posited natally is any area in which you need security, clarity and dependability. As transiting Neptune activates natal Saturn, you could feel generally disoriented and as though your foundation were crumbling, particularly in terms of the house which holds natal Saturn. Neptune, however, is not usually suggesting that you dissolve your entire structure. Flaws are pointed up so that you can eliminate them and improve your situation. Work on what is not perfect and hold onto what seems right. Avoid the temptation to give up totally and let chaos reign. There are always choices.

For a concrete example, let us say that Saturn is in the fourth house, indicating that the home is an area in which you need to feel safe and secure. As Neptune enters the house, you are being told that you have about 14 years in which to make your home closer to the ideal. As Neptune conjoins Saturn, you might begin to notice that there are cracks in the walls of your house. They may have been there for years, but this is the time that you become aware of them. Your options are at least threefold. You could have the cracks repaired, and you will have a firmer physical foundation. You could decide that the cracks are unimportant because the house itself was not necessary for your security—rising above the material. Or you could just watch the cracks, worry about your life deteriorating and wait for the house to collapse. In other words, whatever the circumstances with Neptune transits, you could work to make them better, learn to look beyond the material, or allow your self to be confused and/or disgruntled.

With transiting Neptune activating natal **Uranus** we are dealing with qualities less tangible than those of Saturn. How do you physically measure freedom and individuality? The polarities of this combination could be finding the perfect expression of your individuality or creativity on the one hand; and on the other, you could experience guilt or helplessness in asserting your independence. Or you might feel that others are taking advantage of you so that you cannot "do your own thing." Alchemically you could combat the latter by becoming involved in a worthy cause. You could use these energies to champion the underdog. In this way you would altruistically (Neptune) express the revolutionary and assertive characteristics of Uranus for an idealistic reason.

When transiting Neptune aspects **Pluto**, you might feel as though your power were being usurped; or, conversely, you could remove obstacles that keep you from being in control. Neptune represents the dissolving principle, and elimination and transformation are ascribed to Pluto, as transiting Neptune aspects natal Pluto there is also the potential of making positive improvements in your life. It is a time to let go of deep-seated resentments and obsessions that have been detrimental to your development.

Sometimes one begins to move toward transformation after feeling powerless. One form of powerlessness can be illness. This was the case with a man who had natal Pluto in early Cancer and natal Neptune in early Leo. Simultaneously, transiting Neptune opposed his Pluto and quincunxed his Neptune, and transiting Pluto trined his Pluto and squared his Neptune. At that time he became seriously ill and, because of his weakened state, was totally dependent on others. Over the years he had become bitter and mistrustful and allowed very few to become close to him. If someone did something nice for him, he was suspicious and seemed to find reasons to criticize the individual and to explain the ulterior motives of that person. In his helpless state with the Neptune and Pluto interchange one might have expected him to become even more bitter, but what occurred was quite different. He faced old resentments and let go of them. He contacted people he had not seen in years, forgetting how they had supposedly victimized him. He was more at peace than he had been in some time. What was most amazing was that as he lightened his emotional burden, his physical condition improved.

Let me say immediately that it is not necessary to become ill with transiting Neptune aspecting natal Pluto. There are many other ways this combination may be manifested. When you are aware of the principles involved, you can begin to analyze situations. You can then work on dissolving qualities that interfere with your transformation or asserting your power.

As Neptune aspects your **nodes**, you could have a strong urge to help others or you might be concerned about people taking advantage of you. It is possible to form an alliance that is beneficial to you during such periods, but you also could become confused about relationships that previously have been very clear. One woman "on the singles scene" complained that as Neptune squared her nodes, she kept meeting people who either drank too much or were not honest with her. I suggested that she become involved in some spiritual or artistic endeavor with others. She decided to take a painting class and found that although she did not immediately find Prince Charming, she met people who were neither blatant liars not alcoholics and whose company she enjoyed.

With Neptune aspecting your **Part of Fortune**, you might become disenchanted with goals that you once thought would be fulfilling. Or you could feel somewhat fragmented in terms of your life direction. The potential polarity, however, is to determine the ideal goal to attain. These can be propitious times to investigate aesthetic or universal fulfillment. If your aspirations have been solely materially oriented, you will probably be dissatisfied with the results even if you accomplish the heretofore-desired goals. Becoming more attuned to the universe will be more gratifying than accumulating personal rewards. In other words, if you look beyond yourself for what you want, you will probably feel less fragmented and more whole within yourself.

Since the **MC** represents the way in which you interact with the world, as Neptune aspects the MC, Neptunian themes can be externalized. To illustrate this, let us look at the experiences of a young man who over a two-year period had Neptune moving back and forth over his MC. Although the aspect was intermittent, the entire two-year period was rich with Neptunian symbolism. Only after Neptune made its final pass did its manifestations become less dramatic. His life during that time so perfectly epitomized the various facets of Neptune that I promised one day to immortalize him! The first event that coincided with the transit was a physical attack upon him, for no apparent reason. No one witnessed the assault, and the individuals who victimized him were never prosecuted although the authorities seemed to know who they were. As a result of the attack his classmates were sympathetic and went out of their way to be pleasant to him. He had never felt so charismatic. There were a number of other expressions of both sides of Neptune. At one end he was sometimes unclear about his public image and occasionally felt like a "scapegoat" (to a lesser extent than the above example). But, on the other end, his musical talents flourished and he won a statewide singing competition. During that period he also became a magician's apprentice (a positive alchemical application of Neptune) and still entertains friends with his sleight of hand.

Transiting Neptune aspecting the **Ascendant** can materialize, as with the **MC**, in personal charisma. But Neptune to the Ascendant can also be associated with self-doubt, confusion about self or concern with personal victimization. In fact, you could experience both sides con-currently. I have heard individuals under these aspects make such statements as "I can't understand why people are being so nice to me." And in the next breath say, "I don't deserve such treatment" or "I wonder what they really want from me." You can become vaguely aware of your own shortcomings during these periods. If you dwell on these, you will probably not be able to determine precisely what is wrong with you (we are, after all, dealing with Neptune) but you will undoubtedly succeed in making your self more unhappy. It is better to make progress with the good than directly combat evil. Therefore, you should concentrate on the nonmaterial aspects of your character such as altruism, spirituality and artistry.

In several instances, I have found that clients under this combination expressed their feelings on canvas or through writing poetry. This alchemical ritual, they insisted, helped them better define what they were experiencing. One woman, an artist by profession, was in the midst of drug rehabilitation as transiting Neptune conjoined her Ascendant. It was a chaotic period for her, and as a temporary escape she created a series of paintings. She produced most of these pictures, when she was upset or confused. The process seemed to relieve her distress, but as she viewed the finished products, she also better understood her situation. She has never shown these pictures public but keeps them as a reminder of what she went through and of what she wants to avoid in the future. Even if you are not an artist, you might be pleasantly surprised at what a pen or a paintbrush can create during these aspects.

Unlike aspects to planets and points, which describe specific energies and provide relatively short time spans, a planet transiting a house offers a wider range of possibilities. If transiting Neptune were aspecting your Sun, you would have a number of months to work on your ego. But, if Neptune were moving through your sixth house, periodically, during an approximate 14-year period, you might have to deal with issues of one or more facets of that compartment of your life. You could be faced with matters of health, service, daily routine, etc. Therefore, you do not have to wake up each morning for 14 years with Neptune in the sixth house uppermost in your mind, but only when conditions warrant it. You might select the time when Neptune crossed the cusp of that house to initiate some Neptunian actions in terms of that area, but for the most part it is sufficient to be aware of how the symbolism of the planet might emerge, and handle it when it arises.

This is, of course, true of any of the outer planets in transit. With this method you first determine the length of time the planet will be in a particular house. Next, familiarize yourself with both the positive and negative qualities ascribed to the planets so that you will know which planet is demanding attention as situations arise. If the manifestations are to your liking, you are probably doing something right. You can feel assured that you are moving in the proper direction, and you can sit back and enjoy it. However, if your life is not going as you would want it to, you can do something about it. When issues involve pettiness, victimization, imperfection (especially in situations that have previously seemed ideal), confusion or chaos, Neptune is usually on the scene.

After you have identified the planet associated with the symbolism of the difficulty, you should then try to rectify your position in terms of **that** planet. You should not, for instance, try to exert the physical force of Pluto with Neptune energies. Still, you have a great deal of latitude. With Neptune you could allow yourself to be used and abused because you think you deserve it, and just feel sorry for yourself. That, obviously, would not be the best reaction. Instead you should examine the unpleasant circumstances in order to weigh then in the total scheme of things. If they are intolerable, you can dissolve them. If they are relatively unimportant you can ignore them, thus rising above them.

An excellent way to learn about the meaning of planets transiting houses is through hindsight. After a transiting planet has left a house, you can determine which experiences for the proper time frame can be attributed to that planet and that particular area. This is a technique I frequently use in transit classes, and have found it extremely effective. By sharing experiential information with each other, all participants can increase their understanding of the transiting planets and also have reference material that can be used at the appropriate time for themselves or for clients. Specific experiences will, of course, vary, but certain principles will be repeated and will be applicable to many different types of situations.

With Neptune transiting a house, as with its aspects to planets and points, we are interested in themes of victimization, transcendence, perfection and imperfection, altruism, etc., particularly in terms of issues that pertain to the area through which Neptune is moving. This method will be more fully explained in Chapter Ten.

## Alchemy

A number of alchemical applications of Neptunian energies have already been discussed, but it might be helpful to have a comprehensive list to avoid getting lost in the pages of this chapter—especially if the reader is experiencing Neptune transits! I will also be adding a few possibilities that as yet have not been mentioned. They include:

1. listening to music, or better yet, performing it
   —singing or playing an instrument
2. dancing
3. painting or writing
4. watching movies
5. creating illusions, such as with magic tricks
6. making wine—but don't drink too much of it!
7. meditation
8. visualizing—picture your situation the way you would like it to be.
9. volunteering to help others who are worthy
   —to help avoid victimization
10. looking for answers through prayer or with the *I Ching*
11. enjoying nature, especially the ocean

These alchemical rituals can temporarily alleviate feelings of confusion, which include the "cobwebs in the head," and concern with being used by others. They may prevent your succumbing to the negative side of Neptune, and help overcome the temptation to allow yourself to suffer because Neptune says you should. Even though pettiness and unsavory situations might enter the scene, keep in mind that Neptune is really advising you to move toward perfection, to become more at one with the universe; and let that guide you in the right direction. You can become disillusioned and dissatisfied under Neptune aspects and feel as though your world were crumbling. Be aware of these Neptunian pitfalls but don't settle for them. Even if you experience the negative side of Neptune, recognize that you can also find peace and harmony with this planet and, with some effort, may move a little closer to truth and wisdom.

## Transiting Neptune in Houses

a. Neptune in the 1st house
b. Neptune in the 2nd house
c. Neptune in the 3rd house
d. Neptune in the 4th house
e. Neptune in the 5th house
f. Neptune in the 6th house
g. Neptune in the 7th house
h. Neptune in the 8th house
i. Neptune in the 9th house
j. Neptune in the 10th house
k. Neptune in the 11th house
l. Neptune in the 12th house

1. _____ Be distrustful of your partner or undertake spiritual activities with your partner.

2. _____ Your money may seem to magically disappear so it is better to concentrate on spiritual values, and financial matters may fall into place.

3. _____ Be confused about your personal identity or involve yourself in a artistic project, such as water colors to express yourself.

4. _____ Decide that you cannot accomplish anything and just waste your time or be flexible in your daily routine but don't abandon it.

5. _____ Be suspicious of your children or try to have trust and faith in them.

6. _____ Become disillusioned by your religion or get deeply involved in it.

7._____ Could be risky for investments or handling other people's finances, but good for getting involved in understanding psychological motivations.

8._____ Complain that no one understands what you are saying, write fiction, or communicate on spiritual subjects.

9._____ Allow yourself to be used professionally, or selectively help others through your career.

10._____ Become mistrustful of anything you cannot see, learn to interpret dreams, or develop belief in an occult subject.

11._____ Feel like a victim or outcast in a group or join a spiritual or artistically oriented group.

12._____ Allow your environment to disintegrate or become chaotic or relax in the home and think of it as your sanctuary.

## Transiting Neptune in Aspect

a. Neptune-Sun       f. Neptune-Jupiter    k. Neptune-MC
b. Neptune-Moon      g. Neptune-Saturn     l. Neptune-Ascendant
c. Neptune-Mercury   h. Neptune-Uranus     m. Neptune-Part of Fortune
d. Neptune-Venus     i. Neptune-Neptune    n. Neptune-nodes
e. Neptune-Mars      j. Neptune-Pluto

1._____ Feel misunderstood in everything you say, discuss spiritual subjects, or write fiction.

2._____ Feel victimized in relationships or seek out spiritual or unselfish people.

3._____ Suffer in love or do something artistic.

4._____ Decide that you are ineffective in expressing your will or work on spiritual gratification.

5._____ Try to figure out what people are trying to get from you while only pretending to be attracted to you or enjoy the personal charisma that you have.

6._____Decide that you don't know what fulfillment means or do something altruistic for personal fulfillment.

7._____Allow your power to dissipate, use it to take advantage of others, or use your power to help others.

8._____Totally dissolve your life structure or work on dreams that fit into your life structure.

9._____ Complain about being victimized in your profession or selectively use your career to help others.

10._____You could allow your emotions to confuse you or you may be highly intuitive.

11._____ Do nothing because you are too exhausted or take the initiative in spiritual or artistic matters.

12._____ Remain in a state of confusion, blend polarities within yourself, or become more highly evolved.

13._____ Do a lot of drinking or develop through spirituality.

14._____ Decide that you cannot possibly be independent or use your original ideas in art or religion.

# Chapter Eight

# Pluto

Pluto completes a cycle in 248.4 years. Its velocity is the most inconstant of all the major planets in that it takes as few as 12 years to transit some signs and as many as 29 years to travel through others. Because of its varying velocity through the zodiac, its cyclical aspects do not correspond with particular ages. People born in the late 19th century experienced Pluto square Pluto as late as age 75; those born in the 1920's in their fifties; and those born in the 1960's will have it as early as age 37.

Since about 1940 Pluto has been moving at approximately the same velocity as Neptune, and these planets have been in sextile to each other since that time. Therefore the significance of Pluto transits lasts as long as those of Neptune. Because of retrogradation, the outermost planets frequently activate natal planets and points three to five times.

## Function

In mythology, Pluto was the god of the underworld and in astrology Pluto is associated with retrieving things from the depths. So it is not surprising that it is connected with psychotherapy, nor that subterranean eruptions are ascribed to it. The positive functions of transiting Pluto include delving deeply into matters in order to eliminate nonessentials and thereby making room for transformation. Power, too, is a Pluto motif, and you could find that power figures or matters involving your own power are prominent during Plutonian periods. Plutonian issues cannot be ignored; if you attempt to ignore them, you could experience the "subterranean eruptions."

## How You Might Feel with Pluto Transits

With Pluto there is frequently the feeling that a great change is about to occur. This can be accompanied by a desire to eliminate much of your present and past and start totally anew. There can also be, however, an inkling that such important changes will be irreversible, and therefore there is a hesitancy to make immediate and drastic changes such as those associated with Uranus.

The difference between the type of changes ascribed to Uranus and those of Pluto was made very clear to me by a group of people born in 1952 and 1953 with the conjunction of Neptune and Saturn in their horoscopes. As mentioned earlier, a series of clients with this natal aspect appeared when transiting Uranus was conjoining this conjunction. At this time rapid and dramatic changes were often made without advanced planning or thoughts of consequences. Several years later, as transiting Pluto approached the point of the conjunction, a number of these individuals reappeared. Again there was the feeling that change was imminent, but the urgency of Uranus was replaced by unrelenting but slow-moving Pluto. There was a need to examine all ramifications before a move took place. The natal charts belonged to the same individuals who had experienced the Uranus transit, but the attitude was very different because transiting Pluto instead of Uranus was involved.

One young woman, who was a nurse, was thinking about going back to school to become a physician. When she came in, her main worry was that some golden opportunity might pass her by if she did not act soon. Yet there was the feeling that she should look into all possibilities and make her decision carefully. Together we mapped out a timetable during the Pluto transit by which she could determine what opportunities were available and still be able to register for school in the fall if she decided to do this. Accepting that she **did** have time to ponder her decision made her feel less anxious and she could more calmly analyze the total situation. She finally went back to school and was certain when she did that it was the right decision.

During Pluto transits a tendency toward obsession can exist. Certain subjects and ideas will keep entering your consciousness involuntarily, even if you try to suppress them. Or perhaps it is more correct to say **especially** if you try to suppress them. Although you do not want to become so preoccupied with one segment of your life that other areas begin to deteriorate, it is best to examine the troublesome subject. You could expend more energy by trying to ignore difficulties than by facing them directly. Also, if you try to repress problems during these periods, violent eruptions or feelings of powerlessness could occur. You should investigate and analyze the cause of the problems and take steps to alter the situation.

The main reason for trying to pretend that Pluto aspects do not exist is fear of total change that could lead to disorientation and even a sense of identity-loss. When we accept the realization that "disorientation" is not necessarily a key word for Pluto, and that transformation need not take place immediately, Plutonian symbolism becomes easier to deal with. What Pluto suggests is that you slowly uncover and peel away layers until you get to the bottom of things. You should rid yourself of extraneous or detrimental factors as they appear or are unearthed, and build on what remains. By using this approach, Pluto can help you get to the core of self, and basic needs can become very clear. Since you will undoubtedly be intensely analyzing situations and people during Pluto periods anyway, you might just as well apply this inclination to crucial issues.

There is also a physical manifestation that can coincide with Pluto aspects. I have heard it described as the feeling that a bulldozer has run over your body. This is its extreme form, but at least dull aches and pains may occur with Pluto. These usually emerge when you are not trying to direct its energies. The physical discomfort should be considered a warning that you should try to harness the energy Pluto signifies and wield your own power. When you start to do this, the symptoms often abate and seem to magically disappear.

The gamut of Pluto, as already depicted, seems to range from subtle, internal rumblings to volcanic eruptions, the latter materializing when the former are not heeded. Although often both the positive and negative sides surface during Pluto transits, it is possible to evoke the power and forcefulness associated with the planet without necessarily unleashing volatility or devastation.

One other sensation that can be associated with Pluto is that of loneliness. You could have the idea that no one can possibly understand what you are enduring. If you are having problems, you feel that you must work them out yourself. Simultaneously with Pluto transits, as though to prove the point, you may find that when you are in a group, voices will begin to fade out and you will feel as though you are totally alone. This is not a universal experience, but has been told to me often enough to make it worth mentioning. If this occurs when transiting Pluto is activating your natal chart, you are probably not losing your hearing or your powers of concentration on a permanent basis. It is connected with the need to look inward. You need not avoid social interaction. The possibility is discussed to merely allay fears if it should happen to you.

In spite of the facts that you feel alone, and that you have to solve your own difficulties, Pluto is also connected with depth psychotherapy. Therefore, it is very common for one to consider or actually go into psychological counseling when Pluto enters the scene. There seems to be an awareness that the deep-probing, which is necessary, could be facilitated by an objective person. This is for the purpose of insight. You still ultimately have to solve the problems yourself.

## How Pluto Transits May Be Used

Obviously, you can probe deeply during Pluto transits, whether you do it alone or through counseling. Pluto aspects that are most frequently occurring when one considers or seeks professional help are those to the Sun, Moon or Ascendant. The reason behind this is probably that these three factors represent the core of the person, and if a transformation is going to occur at a primary level, investigation should be thorough. This does not mean that it is essential to undergo psychotherapy during these transits. Nor does it indicate that you cannot seek help at other times. It is just one way in which these aspects can materialize. As we go into each planet and point, alternatives will be discussed.

The power principle, too, will come to the fore during Pluto transits. You will be more sensitive to such issues, whether or not these matters are prominent in the natal chart. One woman had transiting Pluto aspecting her natal Mars in her sixth house. Ordinarily she had no difficultly in taking orders, but now she felt that her boss was bullying her. She became very angry and began to verbally lash back at him. Since this was totally uncharacteristic of her, she was uncomfortable with the feelings. Alternatives to the anger would be to find positive outlets such as competitive sports for the powerful physical energy represented by transiting Pluto aspecting natal Mars.

Since the velocity of Pluto is so inconstant, Pluto square Pluto (the only hard Pluto cycle aspect I have observed) cannot be associated with a particular age or ages. But whenever it occurs, certain symptoms seem to be standard. Deep, important changes are usually being considered and frequently are eventually made. The examples I have collected include, among others, moving to a new locality, following a different lifestyle, and changing jobs or professions. Change of location ranges from moving a distance from where you previously lived (such as one client who left New York and settled in Florida) to buying a home in the same city in which you heretofore rented. The lifestyle examples vary from acquiring a new group of friends and leaving the old behind, to completely changing the pace of life as exemplified by the client mentioned above. She not only moved from New York to Florida, but did this mainly to get away from the hectic pace of New York living. The career category also contains many different types of changes. At the one end, more power in the present job or simply a move from one firm to another within the same profession is possible. At the other extreme, totally changing one's field could occur, such as one client who left real estate to become an antique dealer. It is possible, too, that there may be internal reorganization rather than overt action.

The Plutonian suggestion to eliminate or transform sometimes takes place quite easily, especially if you are cognizant of its presence and are handling it directly. However, if you do not take the initiative you could get pushed quite strongly to take some action during Pluto square Pluto, or for that matter, with any Pluto aspect. One woman experiencing Pluto square Pluto came for a consultation after she had lost her job under very unpleasant circumstances. Her natal Pluto is in the tenth house, and transiting Pluto was squaring it from the first house at the time of the firing. She was devastated by the event because she was made to feel stupid and incompetent. Intellectually she knew that it was a personality conflict between her and her boss, but emotionally she was crushed by the treatment she received from this woman.

As we discussed the needs in her career as depicted in the natal chart, she said that this was not descriptive of the job she had had, and she also admitted that she had not been happy in that position. She had considered quitting and looking for a new job before she was fired but decided against it because the pay check was important to her and she was afraid that she might not find a new position. I explained the implications of Pluto, and suggested that, since the job had not satisfied her needs, perhaps she was being pushed so that she would find more appropriate employment. This was probably a blessing in disguise, and if she viewed it in this manner, it could facilitate her finding a new position. This was easy for me to say but not so simple for her to accept. Her self-confidence had been greatly shaken and she had to force herself to look for a new job. When she did this, she gained momentum. This seems to be true of any Pluto aspect if you help it along. It may also happen if you do not try to use it. The difference is that if you do not attempt to channel the energy, its forceful quality may not lead to the most desirable outcome. This client did find a position better suited to her and wondered why she waited so long to make her move. Maybe, if she had heeded Pluto sooner, she might have avoided some of the unpleasantness she had experienced at the onset of the aspect. But then hindsight always works well!

Directly facing Pluto may not totally eliminate the negative manifestations, but it does help. I have found that the best approach to transiting Pluto is to first note the planet or point that it is aspecting because that is the facet of one's character which needs to be analyzed and/or transformed. Events with the appropriate symbolism will probably occur during the aspect in question. Some of these experiences will be pleasant and others may be unpleasant. The positive manifestations will reinforce certain qualities. These you will want to retain. The negative occurrences will point up qualities that need to be changed or eliminated. Since Pluto is so slow-moving, you will have ample time to ascertain the accuracy of your analysis. You should neither give up trying, nor make sweeping changes after a single experience. As each combination is discussed below, the concepts of power, transformation and analysis will be covered in connection with the corresponding facet of character.

Pluto aspecting your **Sun** places the spotlight on your ego. You could become aware of the power you wield or of your desire to have greater power. You could exert control over others and/or believe that someone else is challenging your position. The feeling of exposure could make you want to withdraw, but this combination does not readily allow you to remain in the shadows. If you try to retreat, you could be pursued. The strong desire to probe deeply into your own ego needs can assist you in sorting out information that comes into your life during Pluto aspects. As with Saturn, Pluto manifestations are often quite explicit. You can experience the heights and depths under Pluto. With transiting Pluto activating your Sun, you can attain recognition and a sense of power on the one hand, and, on the other, bad notoriety or feeling of being controlled by others. As you analyze what is happening, the positive experiences will be highlighting the attributes on which you should build. The negative will point up the qualities that should be changed or eliminated.

When Pluto was squaring his Sun, a business executive was given a promotion, which placed him in charge of a number of people. He liked having the authority and the recognition, but certain side effects began to emerge. Former colleagues were now subordinates and no longer treated him "as one of the guys." The former camaraderie was replaced with respect by some and with animosity by others. He felt very much alone. In fact, when I saw him shortly after the promotion his first words to me were, "The old cliché is really right. It is lonely at the top." He became extremely sensitive to the reactions of others and found that he was painstakingly analyzing every interaction he had. He is still not completely comfortable in his new job, but is getting to know some of the people who are on new managerial level. He feels less isolated, and that, coupled with the power and ego gratification is compensating for what he lost.

Pluto to **Moon** aspects will focus on your emotions or possibly on your subconscious. This is why individuals sometimes begin analysis during these aspects. But, whether you do it with the help of a therapist or not, you could derive insights into your inner self. There is the urge both to understand and to be in control of your feelings. You should be wary, however, of continually holding emotions back, particularly without simultaneously analyzing them. This can lead to emotional explosions. If someone triggers fear, anger or other strong reactions within you, try to determine why.

A woman prone to periodic emotional outbursts discovered, while Pluto was conjuncting her Moon, that her responses were based on a childhood pattern that could be traced back to her mother's treatment of her. Her mother had been dictatorial (the native's natal Pluto was square her natal Moon) and she had rarely been allowed to show her feelings. Thus, as an adult she inwardly reacted strongly against authoritarian types. She attempted to contain her feelings. This worked to a point, but sometimes ultimately led to emotional outbursts which were not always in her best interest. With astrological and psychological counseling (which began when Pluto first conjoined her Moon), and concentrated personal effort she is learning to analyze and express her feelings at an early stage in a situation. In other words, she is giving herself permission to let out her emotions.

When transiting Pluto aspects natal **Mercury**, you could find that your words are heard whether you want them to be or not. But it is also possible that your internal thinking processes are broadened and deepened. If there are important issues you wish to discuss, prepare and rehearse what you want to say before you actually speak. It is not easy to retract statements made under these aspects; and if you speak unwisely, your words might haunt you. You could become so obsessed with what has transpired that you play the scene over and over again in your mind. Recognizing that your words could have a powerful impact should help you to be cautious (but not muted) during Pluto to Mercury aspects. These periods can be very positively used for research. This could be considered an alchemical application in that you are keeping occupied and thereby possibly avoiding rash communications or continual reliving of scenarios that cannot be altered. But it also can produce direct results. As Pluto squared his Mercury, a writer found important background data for a piece he was writing that had previously eluded him. You may not be an author, and, therefore, would not want to spend hours going through books in a library. However, any mental activity pursued during these periods could be investigated in great depth and add a new dimension to your life.

As Pluto aspects your **Venus**, it could indicate the potential of a powerful love relationship. If you are unattached, you might select such a period to place yourself in a situation where you could meet someone of interest. If you just stay at home, it is highly unlikely that "Mr. or Ms. Right" will come knocking at your door, and the aspect will probably manifest itself in some other way. There are a number of alternatives. In existing love relationships, power or possessiveness could be an issue, analysis could take place, and/or the relationship could be transformed or terminated. Then, of course, Venus might also be expressed through sociability or art. If difficulties in love relationships surface at this time, try to look at it as an opportunity to better understand your basic needs as far as love and affection are concerned. While you are examining your requirements, you can also utilize some of the less serious implications of the combination such as social interaction or involving yourself in the arts. This can temporarily alleviate some of the intensity of Pluto without abandoning your direction. You might also select Pluto to Venus contacts to face issues of self-indulgence versus self-mastery, such as overeating, smoking, drinking, etc.

A woman with transiting Pluto aspecting her natal Venus separated from her husband and became obsessed with her situation. She started personal therapy in hopes of better understanding her circumstances in order to change her attitude and her life. She had withdrawn from her friends, and in her solitude was becoming more and more depressed. I suggested that she keep pursuing deeper understanding with her analyst, but that she should also force herself to be with people even for short periods of time. She did this and found that it gave her some relief.

Pluto to **Mars** aspects more than any other combination, can be associated with violent eruptions. Trying to ignore them is like attempting to place a cap on an active volcano. There is a tremendous amount of energy connected with Pluto to Mars, and if you do not select ways of using this force, you might attract people who will use it for you, or on you. If you analyze how you would like to initiate or be assertive and begin to take action, you could take giant strides forward. Mars takes on great momentum when it is activated by Pluto—momentum or power that you should apply to something productive rather than project onto others. This would be an excellent time for highly competitive pursuits, which allow you to focus power on winning. Or it could be a good period for deeply gratifying sexual needs.

When Pluto aspects **Jupiter**, you can expect "a lot"—a lot of something or a lot of nothing. This point was made clear to me by two electional charts that marked the beginning of two businesses. The charts were erected for the same day at the same time, for two locations that were just a few miles from each other. Both had a Pluto-Jupiter conjunction in the second house. This time was chosen because making money was one of the main concerns of each person. Since these charts were almost identical, you would expect the end result to have been the same, but this was not the case. One business has flourished and brought in great revenues; the other went bankrupt. We might find some reasons for this in the natal horoscopes of the persons involved, but it still serves to point up the extremes of the Pluto-Jupiter combination. In the case of the successful business, the man had apprenticed in the field in which he had established his firm and thoroughly examined the need for the services he would offer. In the other example little forethought was given to the manner in which expenditures were made. The electional chart predicted riches; therefore, the second man felt it safe to make large purchases. He overextended himself financially on inventory, without determining if the demand warranted the supply. Although these examples come from electional charts, they not only describe the extremes of Pluto-Jupiter, but also provide clues as to how the individual can beneficially deal with Pluto transits to natal Jupiter. Jupiter indicates the need to grow, and transiting Pluto aspecting natal Jupiter states that you might want to develop in some new way or to move in a different direction. To avoid the excessive or extravagant side of Jupiter, apply Pluto's thorough and deep-probing analytical ability, and you will undoubtedly make noticeable progress. The man with the successful business mentioned above will attest to that!

As transiting Pluto activates natal **Saturn**, you may be called upon to transform some part of your life structure, or to analyze and possibly change certain responsibilities. Since power is ascribed to Pluto and authority to Saturn, issues involving such matters might also emerge with these aspects. You could experience a psychological heaviness from the weight of your responsibilities, or you might feel depressed and even oppressed during these periods. If, however, you analyze your position in regard to your environment, your duties and need for control, as these aspects begin to form, this will not only relieve the negative feelings, but also could assist in clearing defining actions that can then to taken. This combination, more than any other, can make issues crystal-clear. Pluto offers the opportunity to get back to the basics, which Saturn represents.

A dramatic illustration was presented to me by a woman who, when making her initial appointment, informed me that her husband had left her, she was having serious problems with her children, and her sister with whom she had previously been very close had broken off contact with her. "In other words," she said, "My entire life is falling apart!" And she concluded with "My world is disintegrating and I feel totally powerless." It certainly sounded Plutonian, and when I examined her horoscope and transits, I found that she had transiting Pluto moving through her third house (siblings), squaring Saturn (life structure) in the fifth house (children); and Saturn co-ruled her seventh house (partners).

Since the third house is also the house of communications, and Pluto advises you to bring deep matters to the surface, it seemed logical that she should talk to the people with whom she was having difficulty. I explained, however, that Pluto aspecting Saturn did mean her life structure should, and could, be transformed in some way and this might indicate that some of these relationships could end if she took action. She was so unhappy that she decided any change was preferable to the present state of affairs. Although all the relationships were important, she started with her sister who was least crucial to her. She was leery of a face-to-face confrontation and, therefore, wrote a letter describing her own feelings and asking her sister what was wrong. Once the issues were brought into the open, the situation improved. This gave her courage to deal more overtly with her children and her husband. In every instance there was a change and an improvement.

Pluto to **Uranus** aspects are probably the best to have when you want to start a revolution. Symbolic of this were the riots, which took place in some of our cities in the mid-sixties when transiting Pluto and Uranus were conjunct. By expressing your individuality and creativity during these aspects you could avoid the potential violence caused by the suppression of your self-expression. If your position is such that you are unable or fearful of directly asserting yourself with the individuals that seem to epitomize the aspect, there are alchemical ways to activate the energies. A newlywed believed that her mother-in-law was trying to destroy her marriage. Among the transits she was experiencing at that time was Pluto square her natal Uranus. She had the strong urge to tell the woman off but felt too insecure in her new role as wife to do this. Instead she became actively involved in a political cause and was elected spokesperson for a local group. This did not thwart her mother-in-law's efforts to undermine her, but it did add to her husband's admiration of her and strengthened their relationship.

When transiting Pluto activates natal **Neptune**, your religious or spiritual views could be transformed. A young man who natally has Neptune sextile Pluto (an aspect common to everyone born since the early forties) forming a yod with his MC went to live in an ashram when Pluto conjoined his Neptune. He had been raised Catholic. With the first pass of Pluto he had begun to question his religion and started to investigate alternatives. With the third pass he made his move.

During such aspects there can also be a heightened awareness of situations that are unclear, or ones in which you might feel used and abused. Analyzing and facing such situations can be very effective. You can clear the air by recognizing and eliminating obstacles. And with supposed victimization, by bringing these feelings into the open you might improve your position or discover that you have misread the motivations of the other person. Neptune is, after all, imagination. Therefore it is possible to create fiction with this planet and become obsessed by your creation when Pluto aspects it. It is undoubtedly preferable to probe into the basis for these feelings to determine their validity rather than allow resentments, which may be unfounded to build to the point of explosion.

As transiting Pluto aspects your **nodes**, you could find yourself coming into contact with powerful people, or power may become an issue in a present association. Be aware of this and be prepared to stand up for your rights. It is also possible that a relationship could be transformed or end. If during these periods, you try to determine what you want and need from others, it will probably make it easier to express your requirements clearly or to deal with relationship matters that might arise. If you are establishing your expectations, it will be relatively simple to see whether or not another individual is living up to your standards.

With transiting Pluto aspecting your **Part of Fortune** you could feel a powerful sense of fulfillment, or perhaps face the realization that a goal toward which you have been striving is not attainable or no longer desirable. I distinctly remember thinking that I had finally found myself, and my direction when Pluto conjoined my Part of Fortune. It was a period in which I was enjoying my teaching and when I wrote my first book. My Part of Fortune is in the third house, and it was at that time in my life that

I became aware of how important communicating with the world at large was to me. If you become dissatisfied with your goals under a Pluto transit to your natal Part of Fortune, you should heed the warning but control the urge to discard everything. By analyzing your position you could discover that some elements are salvageable, and you need not go back to "square one." It may not be the entire project that should be eliminated, but merely one segment of it. Or perhaps only the methodology need be changed. It would be a shame to discard anything that could later prove to be valuable. Examine step-by-step where you are, particularly in terms of the house in which the natal Part of Fortune is located, and this could provide enlightenment.

Transiting Pluto to the natal **MC** could herald a totally new profession or a different public image. Keep in mind, however, that this need not occur overnight. Not only do you have 3 to 5 passes of Pluto to the MC which could take one to two years, but you also have the 12 or more years it will take Pluto to move through the tenth house, to make necessary changes. If sweeping alterations need to be made, you can do it slowly, deliberately, after thorough investigation. Whether or not transformation is an issue, power, too could be a concern. You might be given more power in your career or become aware of others wielding authority over you. You may want to flex your muscles and let people know that you do not want to be ruled by them. Find positive ways to assert yourself without running roughshod over others. One student told me that this was the period in which she did not try to boss others around but did learn to say "no" to them.

When transiting Pluto activates the **Ascendant**, as with the MC there can be emphasis on power, but from the perspective of self-control rather than control over others, or in the world generally. You might also want to change yourself in some way, and this can manifest itself in dressing differently—not necessarily bizarrely, but in a style that previously was not appealing to you. It is fascinating to observe how outward appearance can reflect what is happening internally. Since Pluto to the Ascendant can alter the way you look, it can be a propitious time to lose or gain weight. It is essential to "psych" yourself properly if you want to diet because the obsessive, unrelenting quality of Pluto will help you follow through in one direction or the other. It is possible both to go on eating binges, and to take pride in self-denial during these aspects. If the opposite of what you want is occurring, you will have to work on brainwashing yourself.

The **conjunction** of Pluto to the Ascendant more than any other aspect is the one that most consistently coincides with externalization. After having spent a number of years looking inward as Pluto traveled through the twelfth house, there is a need to resolve matters that have been analyzed but not yet faced. As transiting Pluto conjoined her Ascendant, a woman believed it necessary to make a choice between her husband and her lover. She had been having a secret affair for several years and had coped with both relationships, but at this point she felt she had to make a decision as to which man she wanted to spend the rest of her life with. No one was pressuring her. It was coming from within. She said that she had been thinking about it for some time, but it was not until Pluto moved from the twelfth house into the first that she felt compelled to take action. Since Pluto was entering her house of personality, she not only felt that she had to bring her relationships out into the open, but she also wanted to transform herself. The way she chose to begin the process was to have an astrological consultation for her self and then a relationship reading with each of two men. At the same time, she started marriage counseling with her husband. She did not rush into making a decision. Instead, she brought problems into the open and analyzed them thoroughly before she took definitive action. Ultimately, she and her husband worked out their differences and she stayed with him, having been convinced that it was the right decision.

Pluto moving though houses should be handled in much the same manner as one copes with aspects from transiting Pluto to natal planets and points. It should be dealt with in terms of analysis, power, elimination and transformation. The time span for Plutonian changes in areas of one's life is, of course, much longer than with single aspects. When Pluto conjuncts the house cusp, a situation or two may arise that fits the symbolism of Pluto and the specific house. When Pluto conjoined my IC, our last child was married (we selected neither the date nor the time). And the home environment was greatly altered when he left. I also chose to have our house painted at the same time. I did this for the unastrological reason that some of the activities connected with the wedding were being held at our home. Astrologically, however, I knew that changing the appearance of the house would be a positive way to use the Plutonian energies. We won't discuss the power struggles with the painters! Suffice it to say that the end result was worth the battles.

After the initial entry of a planet into a house, you will not constantly be aware of its presence during its entire stay there, nor should you be. Periodically conditions will warrant Plutonian probing or transformation that is appropriate for the area involved. Or sometimes, without external stimulation, you will feel you want to make changes there. Keeping in mind the house that Pluto, or any other planet, is transiting will help you more quickly and positively deal with issues that might arise, or assist you in making conscious use of the energies symbolized.

## Alchemy

Plutonian issues have the capability of being more devastating than those of the other planets. Unlike the changes evoked with Neptune and Uranus, which can and should initially be tentative, moves made with Pluto were usually more sweeping and, therefore, more difficult to reverse. Yet, of all the planets, its alchemical rituals seem to work most easily. So, if you are having Pluto aspects, and feel that grand changes are imminent, but are not sure what action to take, don't hide your head in the sand. I have yet to see that approach bring satisfactory results. Start by cleaning out closets and drawers. You need not tackle the entire task at once nor do you have to throw everything away. You can start slowly (perhaps examining only two garments a day if the prospect of an entire closet is overwhelming) until the task is completed. Only get rid of things that are no longer necessary or useful. Literally making more room, for some reason, seems to make more space in your life in other ways. Often even people who know no astrology but are having Pluto aspects will perform this ritual by instinct before they are advised to do so. The woman mentioned above, having the problem with the husband and the lover, arrived fifteen minutes late for our first appointment because she had felt a compulsive urge to finish cleaning out a closet before she left home. I have many other such examples. In fact, it was after about four or five people told me similar stories while transiting Pluto was in the picture, that I began to use cleaning closets and drawers as my principal alchemical ritual for that planet.

If you are a super Virgo, and are so neat that there are no belongings to be culled from your closets and drawers, you can substitute taking out the garbage. This ritual, too, if effective. You might notice that garbage seems to pile up more quickly during Pluto transits. This is a sure sign that something needs to be unearthed and eliminated.

The final ritual that should be mentioned is that of restoration. Not everyone, of course, is in a position to renovate a house. Nor does every house need renovating. But if it is applicable, by all means use it. Then, too, there are other objects such as furniture that can be reupholstered or refinished. I particularly like this application of Pluto because it exemplifies the idea that Pluto does not mean total transformation, but rather building on a strong foundation. For instance, if you are refinishing a chair, you might have to strip away layer after layer of paint until you get down to the natural wood. You have not changed the size or shape of the chair. You have just gotten it back to its original state. Then you can refinish it in any manner you wish.

Keeping this illustration in mind as you experience Pluto might help to relieve some of the anxiety that can accompany Pluto aspects. Pluto does not negate all that you have been and done. Rather it can put you in touch with "the real you" which may have become masked by "layers of paint." It is upon the core of self that you can develop "the new you."

## Transiting Pluto in Houses

a. Pluto in the 1st house
b. Pluto in the 2nd house
c. Pluto in the 3rd house
d. Pluto in the 4th house
e. Pluto in the 5th house
f. Pluto in the 6th house
g. Pluto in the 7th house
h. Pluto in the 8th house
i. Pluto in the 9th house
j. Pluto in the 10th house
k. Pluto in the 11th house
l. Pluto in the 12th house

1. _____You might feel that other people boss you around professionally, you could be given great power in your career, or totally change your profession.

2. _____ Have power problems with coworkers, totally change your routine, or let cleaning away rubbish become part of your daily work.

3. _____ Monopolize conversations, or analyze words deeply before conversing so that you can communicate powerfully.

4. _____ You might feel intimidated by your peers or you could get a position of power in an organization, or be instrumental in transforming a group.

5. _____ Be overly authoritarian, or personally transform and exert control over your own life.

6. _____ End a relationship for no reason or examine carefully any problems that arise in close one-to-one relationships and possibly transform them.

7._____ Feel powerless, probe deeply into the subconscious mind, or wield power behind the scenes.

8._____ Spend all your money and then some, get involved in a project that could bring you great financial reward, or reassess your value system.

9._____ Feel manipulated in your religion or transform through it.

10._____ Let others control you in the home area, take charge of the household or renovate your home.

11._____ You could feel controlled by others because of financial obligations to them, or you may find yourself in charge of someone else's money, or you could probe deeply into your sexual involvements.

12._____ Be dictatorial with your children, deeply analyze them, or take up a creative project with which you can have a powerful impact.

## Transiting Pluto in Aspect

a. Pluto-Sun
b. Pluto-Moon
c. Pluto-Mercury
d. Pluto-Venus
e. Pluto-Mars
f. Pluto-Jupiter
g. Pluto-Saturn
h. Pluto-Uranus
i. Pluto-Neptune
j. Pluto-Pluto
k. Pluto-MC
l. Pluto-Ascendant
m. Pluto-Part of Fortune
n. Pluto-nodes

1._____ Tell people off no matter what the consequences, or you could get your point across effectively in conversation.

2._____ Power could become an issue in your career or you may be transformed through your job.

3._____ You can declare war to get what you want, or you can take the initiative in a less drastic manner to attain power.

4._____ Push everyone around or be more positively influential in relationships.

5._____Cram your spiritual views down someone else's throat or deeply analyze and perhaps change them.

6._____Buckle under and let someone else control you, or powerfully work toward ego gratification.

7._____React precipitately whenever your individuality seems threatened or effectively utilize your original ideas.

8._____Let someone else control you, or take charge of yourself.

9._____Manipulate others through the emotions, or analyze your emotional needs.

10._____ Ignore the urge to change so that you can feel frustrated, or transform some part of your life structure.

11._____Get obsessed with fulfillment and move ahead at any cost, or think deeply about what will give you a sense of fulfillment.

12._____Put your head in the sand and hope Pluto will go away or select some way in which you would like to transform and move in that direction.

13._____Do nothing because the possibilities are overwhelming, or make great strides in your development and growth.

14._____Become jealous and obsessed with a love relationship or deeply analyze it.

Chapter Nine

☊ ☋

# The Moon's Nodes

The **mean** nodes go steadily backwards at a rate of 3 to 4 minutes a day. They move a degree every 18+ days, and are in aspect for about 38 days. The **true** nodes change direction frequently. This chapter is not a stand for the mean nodes versus the true nodes. However the mean nodes are in range for longer periods than aspects of the true nodes. And there is more time to grasp and work with their implications.

## Function

Like the planets, the transiting mean node also has a function. The nodal axis is connected with relationships. In the natal chart the signs and houses in which the nodes are posited, along with the aspects they form with the natal planets and points, provide clues as to how we interact with others. The transiting nodal axis informs us of what others are activating within us at a particular time.

Whether a transiting planet is aspecting your natal nodes, or the transiting nodes are aspecting a natal planet, other people will be playing an important role in current situations. The difference is that if a transiting planet is activating your natal nodes, **you** will be applying the energies ascribed to the planet in your associations. Whereas the transiting nodes aspecting a natal planet indicate that someone else will be triggering the principles associated with that planet within you. For example, transiting Saturn aspecting your nodes could mean that you are examining responsibilities or commitments in a relationship. The idea to do this comes from you. You are the initiator. If, however, the transiting nodal axis were aspecting your natal Saturn, another person is somehow influencing you to examine your responsibilities or commitments. The thought or situation is initiated by someone else, but how you respond is still up to you.

## How You Might Feel with Nodal Transits

Perhaps it is inaccurate to say that the transiting nodes by themselves symbolize or elicit certain rumblings within us. What they do is highlight feelings associated with the planet or point they are activating, as illustrated above. The sensations experienced with the nodal axis transiting Saturn were Saturnian. In other words, another individual is responsible for your examining your commitments. Had it been Uranus, someone else would have been brought forth matters pertaining to Uranus, etc. the common denominator is that the stimulus is external when the transiting nodes are involved.

## How Node Aspects May Be Used

In spite of the fact that transits from the nodes indicate that someone is affecting you, it is still possible to use them advantageously. Being aware of the aspects allows you to know what is going to be activated within you so that you can prepare before it occurs, or at least try to direct the situation as relevant factors arise.

The nodal cycle itself can provide strong evidence of the effect others can have upon us. The nodal cycle was 18.61 years and makes a major hard aspect to its natal position every 4.65 years. The premise here is that some relationship would have an impact on the manner in which you interact with others. With an aspect of transiting node to natal node you would not wake up one morning and, for no apparent reason, decide that today you will start dealing with people differently. Instead, outward circumstances would seem to dictate this. A study on the nodal cycle and its significance in the lives of U.S. Presidents was conducted by Ken Negus corroborating the above.[1] He concentrated on the conjunctions and oppositions and found that the transiting north node conjunction or opposition natal north node in most cases

---

1. Ken Negus, "The Moon's Nodes, the Nineteen Year Transit Cycle and the U.S. Presidents," *The Journal of the Astrological Society of Princeton, N.J,,* Inc., Issue # 2

coincided with events important enough to be mentioned in biographical sketches. The events ranged from entering office and getting married on the one hand, to leaving office and death on the other. Obviously some occurrences were more desirable than others, but all inferred relating to people or society-at-large in a different way. This is even applicable in death since that is the cessation of worldly relationships.

Although none of my clients have the impact on the world that U.S. Presidents have, they still usually experience some kind of external change in their relationships because of others, when the transiting north node conjoins and opposes the natal north node. I have also found that squares are significant, and interpret these much as I do the conjunctions and oppositions. The nodal return, however, does seem to be the most significant aspect of the cycle.

A friend who is a craftsperson and teacher found just prior to her nodal return (natal north node in the tenth house) that she began to have problems in her professional relationships. She said she had felt that something was wrong for several weeks prior to the time that the aspect was in range, but dismissed the idea of changing anything until students who had previously come to classes faithfully became erratic in their attendance or stopped coming entirely. Most gave valid reasons, but it was so atypical that she found herself wondering if she had done something to drive them away, or whether her teaching methods were to blame. When she described her situation and I saw the nodal return in her chart, it seemed clear that this aspect defined her circumstances. It was a time to change the way in which she interacted with the world. This was substantiated when she told me that two years earlier she had had similar feelings. Her products were not in demand and her work seemed to come to a halt. It was the first time she had questioned her abilities, and it was during that period she added a new technique to her repertoire and also began to teach her craft. When I checked the ephemeris, it was at that time that the mean north node entered the tenth house.

As we reviewed her total situation, she realized that not all her experiences regarding her craft and her teaching were negative. She was not doubting her talents at this time, but more her methodology and the manner in which she dealt with others. She was facing these issues at the time of the nodal return, but, perhaps, if she had dealt with these matters when they first entered her consciousness, she might have avoided some of the unpleasantness and hurt.

This example made me think that there might be something we can do with node aspects. Even though others may be influencing us to relate differently (as with transiting nodes to natal nodes) or to get in touch with certain functions within ourselves (transiting nodes to natal planets) perhaps a little groundwork or research in advance might be helpful. It could be that we are given clues in advance of the aspect. If we know what is likely to be activated, we can seek out people who can help us express the appropriate motifs positively, or at least prepare us for the types of situations that might arise. If we understand the issues and are channeling our various drives positively, we need not have negative experiences. Since that time I have been using that approach with clients, and most of the feedback has been positive.

The transiting nodes aspecting the **Sun** indicate that your ego will be put on the line in some way. I have several examples of people who received promotions at the time of the aspect, and this was true whether the aspect was hard or soft. I do have one case, however, in which the sextile from the transiting north node to the Sun materialized in the person's getting fired. This came as a blow not only to the client but to me as well because I was absolutely certain he would get a promotion at that time. After all, that was the way it had happened with every other case I had. That made me realize that, although one's ego is highlighted by someone else when the transiting node aspects natal Sun, the end result might not always be desirable. So my advice with these aspects now is to remember that the world will shine a spotlight on you and you should put your best foot forward. You can take one step beyond this. You can consider the kind of recognition that you would like to receive and work in that direction. This need not, by the way, be restricted to ego gratification in career but can be applied to any area in which you might be, or would like to be, noticed.

When the transiting nodes activate your **Moon**, you could be exposed to emotional situations or you might find that your mothering nature is being called upon. During such times you could be more sensitive than usual to the actions of others and, therefore, you will tend to react emotionally and instinctually in relationships. Emotional issues that have been locked away in the subconscious could emerge and put you in touch with feelings from your past. If this occurs, examine them carefully rather than try to repress them because by doing this you could rid yourself of old obsessions. This sensitivity connected with these could also exhibit itself in tenderness and compassion in contracts you have during these aspects.

Node to **Mercury** aspects advise you that communications in some form or another will be significant in relationships. If there are people with whom you would like to communicate, select a time just prior to the nodes aspecting your Mercury to call or write. You might maintain interaction or be contacted at the time of the aspect. For this reason I tell my writer clients to send manuscripts during these periods. There is no guarantee that the results will be positive, but at least the symbolism is appropriate. Since another person is going to be making contact, why not try to make it someone you want to hear from.

When **Venus** is being aspected by the transiting nodes, you should prepare yourself for social interaction. You might plan a party so that you can invite people whose company you enjoy. In this way you can avoid the possibility of feeling that you have wasted your time. It is not an easy period in which to work diligently. Others may interrupt you to join them in some frivolous activity. It might then be difficult to maintain a tight work schedule even if other transits indicate that you should. You could go on a pleasure trip to take full advantages of Venus. If this is not possible, at least rearrange your schedule to allow more time to enjoy your Venus rather than fight it.

Node to **Mars** aspects indicate that you will probably be asked to initiate in some way, or you could find yourself energized by others. Contacts can be stimulating during these aspects, but you might also find that you are quicker to anger, or that people seem to make you argumentative. Since it is not a time for sedentary interaction, you might want to be sure that you are physically active, preferably with others. Expanding your energies in some form of exercise could avert the possibility of outbursts of temper just in case someone is tending to anger you. One client went on a group backpacking trip during a node to Mars aspect. He arrived home tired but had no unpleasant encounters to report.

When the nodes activate your **Jupiter,** you will probably find that people are helping you to develop in some way, or, as with Venus, your time is being wasted by others. This is an excellent period to sign up for a course or go on a trip. You might still feel overwhelmed, but at least you will have taken a step in the right direction. Exposing yourself to situations and individuals that might help you to grow is preferable to just waiting to see what will occur.

A student who could not believe that north node conjunct her Jupiter would bring anything but "wonderfulness" sat back and waited for abundance to come her way. What occurred was that she was given so much to do at work that she had to put in overtime. This meant a little more money, but she would have preferred to leave work on time. When she arrived home, each member of her family needed her for something. She felt that it was nice to be wanted, but her exhaustion made it difficult for her to fully appreciate it. Finally, her social life soared. It seemed that everyone she knew wanted her company and it was difficult for her to say no. She was happy to see the aspect leave and vowed that next time the nodes activated her Jupiter she would be sure to take a vacation for the whole month it was in range!

During node to **Saturn** aspects you will certainly be examining your commitments, responsibilities or life structure because of other people, not because you particularly want to. Or you might feel limited or frustrated by someone. Since there is no inner urge to think about responsibilities, the aspect might catch you unaware and therefore be restricting. If, however, you know it is coming and prepare for it, you could find that it is an auspicious time to stabilize or evaluate your position in a relationship. Or you might even reap rewards through another person.

An illustration of the latter was given me by a woman who was in the midst of a divorce. She was going to have the transiting nodes square her Saturn during the period in which her court date was set. We reviewed a number of Saturnian possibilities, and one that seemed particularly applicable to her circumstances involved the settlement she hoped to get from her husband. I suggested that she consider her needs very carefully so that they would be set in her mind and she could stand firm. She went one step further. Her husband was not being at all cooperative so she determined what she wanted and asked for more. In this way she could seem to make concessions and still get what she wanted, and this is exactly what happened. She was happy with the settlement and her husband thought that he had won.

When the nodes aspect natal **Uranus**, you could experience spontaneity in your relationships. Someone might suddenly enter your life or abruptly end an association with you. Others might contact you to do something on the spur of the moment. Your individuality, creativity or need for freedom might be triggered at this time. This is not the easiest combination under which to stabilize a relationship. It is much better to "hang lose." With the focus on Uranus you would want excitement, so you might try to plan some exciting activity with another person. It probably will not turn out exactly the way you expect it to but that could add to the excitement. As with Uranus to nodes, nodes to Uranus could also indicate that unusual and/or strange people will pop into your life. A man who commuted to work by train found, during the period of the nodes aspecting his Uranus, a very strange assortment of persons approached him each morning as he waited for his train. He must have enjoyed it because he seemed to like recounting the stories bout it. If, however, you prefer to avoid bizarre people or encounters, you could seek out more creative types with whom you would enjoy being in touch.

As the nodes activate your **Neptune**, you might come into contact with someone who could have a strong spiritual impact on you. This is a good period in which to go on a religious retreat or expose your self to workshops or classes on nonmaterial subjects. By being involved in such activities, you might avoid experiencing the deceptive, illusory side of Neptune. Although dreams can come true with Neptune, it is also possible to delude yourself. Be cautious of people offering you opportunities for material gains when the nodes are aspecting your Neptune. During these periods you may be more vulnerable than usual and possibly even attract people who might take advantage of your vulnerability. Therefore, if possible, it is best to concentrate on the nonmaterial side of life when Neptune is highlighted.

My most poignant example of this was provided by a student with Neptune natally in her fifth house. Before studying astrology she had received an offer to purchase land in Florida, which she could not resist. She jumped at the chance to buy; but later, when she visited the property, she discovered that it was under water. When, after she began studying astrology, she looked back in the ephemeris to that "fateful" period, she found that the transiting nodes had been squaring her Neptune at the time. Had she been aware of the transiting nodes squaring her Neptune and the implications of the aspects, she might have gone to see the property before buying it and thus could have avoided her fiasco.

Another possibility with this combination is to feel burdened or taken advantage of by others. Being of service can be a gratifying use of Neptune, but the satisfaction will be erased if you feel that someone has "conned" you into doing it. Before the aspect comes into range examine your altruistic projects. Note why and whom you are helping. This will define your position and give you an excuse to refuse assistance to those whom you would not choose to serve.

Transiting nodes aspecting natal **Pluto** could indicate that you might be given some kind of power or that you may be called upon to exercise your power. Another possibility is that someone could induce you to analyze yourself and perhaps transform in some way. There may seem to be no way to sidestep a situation that is presented to you. And you might feel that you are being pressured to make an irrevocable decision. If you do not face the circumstances directly, someone else may make your decision for you. Although conditions may not be of your making, you can still influence the outcome.

When the transiting nodes activate your **Part of Fortune**, watch for someone bringing you an opportunity or for the possibility of another individual making you feel good about yourself. If you leave it totally to chance, however, an opportunity might rather be snatched from you, or a relationship may cause you to question what you are striving for. Look for ways of attaining fulfillment through some association. If you take the first step by making contacts, you could open doors for yourself. I have several examples of this, but let me mention just two. The first concerned a man with the Part of Fortune in the tenth houses who was dissatisfied with his job. Just prior to the transiting north node opposing his Part of Fortune he sent out resumes for positions he thought would make him happier. During the aspect he received a positive reply to this application. The other example involves a woman who did not sit back and wait for Prince Charming to arrive as the north node conjoined her seventh-house Part of Fortune. With a little prodding from her astrologer she started setting up social engagements with men who seemed desirable. In the first case cited, a new job was found and so far seems satisfactory. In the latter case, it is too recent to know the degree of success, but there have been some interesting nibbles at the bait.

As the transiting nodes aspect your **MC**, it is your career or public image that will be highlighted by others. You could find that there is more professional interaction—more clients or more people contacting you to do business. In the above example of the man who sent out job resumes when the nodes activated his Part of Fortune, the action was appropriate because the Part of Fortune was in his tenth house. But more than any other set of aspects, nodes to MC is the perfect time to make career-connected contacts. Or you could ask for a raise at that time. Be sure you deserve it, because these aspects will not make the increase in salary magically materialize. The only guarantee is that others will be focusing on your career or public image. You might want to seek recognition during this period, but remember that both fame and infamy fall under that category and it is important to present your best self because the world will see what you project.

Transiting nodes to natal **Ascendant** highlight the more personal you. If you are interested in the way others see you, this is a time to look because they will undoubtedly tell you anyway. Since the node axis represents associations generally, and does not distinguish between acquaintances and closer relationships, it could be that even someone whom you do not know well might have a strong effect on you. If at the time of the node to Ascendant aspect, you feel attacked or disapproved of, do not crawl into your cave. You need to be in contact with other people, so consciously look for those who can make you feel good about yourself.

If one of the transiting nodes is **conjoining** the Ascendant, its implications are somewhat different from a **planet** conjoining the Ascendant. Since the mean nodes always move backwards, as a node crosses the Ascendant, it is leaving the first house, not entering it. And it is moving into the twelfth. It is leaving the realm of the observable and moving into the area of the subconscious. This indicates that others could cause you to look deeply within yourself or could have a profound effect on your identity – an effect that could linger subliminally. This is why, during this aspect, it is a good idea to find people who will be supportive of you. In that way you can start the period in which relationships will have a more subtle effect on you (nodes going through the twelfth house) with self-confidence rather than self-doubt.

With a node conjoining any house cusp, it is, of course, leaving that house, not entering it. For instance, after it conjoins the MC, where the emphasis in relationships has been on career, associations will begin to take on the color of ninth-house matters. Actually, since we are dealing with an axis, it would be more accurate to say that as the nodes pass over the MC and IC both third- and ninth house matters will, probably become significant in relationships. It therefore follows that as the nodes enter a pair of houses, although it is through the back door you will probably become aware that the focus in relationships will begin to shift to matters connected with those houses. You might even find some appropriate dramatic occurrence involving others will take place at that time. As with the planets, do not expect every single day to be filled with people activating matters connected with the houses that the nodes are transiting; but you might pay particular attention to the time of their entrance into and exit from each house.

I first examined node conjunctions to house cusps on my own chart. When the north node entered my seventh house I was married; and when it entered my fifth I had my first child. This encouraged me to watch the nodes going through houses and I began to advise clients of the possibilities of the movement in their charts. The feedback has been favorable. The exercise on transiting nodes in houses will offer possibilities of each area. The nodes being opposite each other will always activate houses in pairs, and it may be that the north node moving through a house will have a different impact than the south node. For the sake of simplicity no distinction is made between the nodes, and the houses are discussed singly rather than in pairs.

# Alchemy

The alchemical rituals are aimed principally at helping reform energies within ourselves, and since the nodes represent relationships, it is questionable as to whether you can directly influence others in the same manner.

People do not always choose to follow your script, but possibly, an alchemical technique by which you could reach others is through visualization. Create in your mind the circumstances in the relationship that would be acceptable to you. There is another alternative however, which is more direct. Pay attention to what the transiting nodes are doing to your chart, both by house placement and by aspects to planets and points. Then you can make an effort to deal with others in terms of the planets, points and houses involved. If someone else is going to have some effect on your sense of responsibility (Saturn), need for independence (Uranus), desire for growth (Jupiter), etc., why not help to decide who it will be? Or, consider what issues may be brought to the foreground and examine the potential alternatives so that you will be prepared for what might be coming. It is possible to sit back and wait to see what others are going to do to you, but it will probably be much more gratifying to have some say in the interaction. In this way you might select the right people to contact and avoid potential problems; or at least be ready to deal with issues that someone else may cause you to examine.

## Transiting Nodes in Houses

a. node in the 1st house
b. node in the 2nd house
c. node in the 3rd house
d. node in the 4th house
e. node in the 5th house
f. node in the 6th house
g. node in the 7th house
h. node in the 8th house
i. node in the 9th house
j. node in the 10th house
k. node in the 11th house
l. node in the 12th house

1. _____People may have strong effect on your values or could help you spend your money. Seek out those who could increase your revenue.

2. _____You could have more business contacts than usual, or someone might cause you to examine some role you perform in the world. Try to meet people who can forward your career, or in some other way project yourself into the world.

3. _____You could possibly be sued or you might get married. A good time to look for intimate relationships that you would like to cultivate.

4. _____ Others may tell you how and when to work or you could organize your daily routine and delegate some tasks to others.

5. _____ Romance could come into your life, or focus might be on your children. For preparation be aware of what is happening to

your children, or if you would like romance, expose yourself to situations that might facilitate this.

6. _____ You could be approached to join a group, or someone might tap into your aspirations. You might choose to investigate organizations with which you could become connected, or interact with those who could help you attain your hopes and wishes.

7. _____ Be wary of possible dependency from other people. Siblings or neighbors could become more prominent in your life. Determine how involved you want to be.

8. _____ You might find yourself subtly influenced by others for either good or ill, or people may cause you to look deeply into yourself. A good time to analyze those with whom you relate.

9. _____ You could meet people through travel or higher education, or individuals who are spiritual or philosophical might make contact with you. You might pick this time to take a trip or become active in a religion.

10. _____ You could feel that everyone is focusing attention on you. Certain people may be complimentary, others may be hostile. Both alternatives are possible, so if you have an unpleasant experience with someone, look for another person who will be supportive.

11. _____ Some individual might offer to share his or her resources with you, or someone could make sexual overtures. In either case, you can make the decision as to what and how much you want to accept.

12. _____ Others could influence you in the home or you might feel that the whole world is coming to visit you. Invite people with whom you would like to interact, and maybe others will stay away.

## Transiting Nodes in Aspect

a. node-Sun
b. node-Moon
c. node-Mercury
d. node-Venus
e. node-Mars
f. node-Jupiter
g. node-Saturn
h. node-Uranus
i. node-Neptune
j. node-Pluto
k. node-MC
l. node-ASC
m. node-Part of Fortune
n. node-node

1._____Others may have an impact on your feelings either being emotionally upsetting or emotionally supportive.

2._____You might feel victimized or someone could spiritually inspire you.

3._____You might feel thwarted from attaining a goal or another individual may offer you an opportunity for fulfillment.

4._____Your ego will be important in relationships. Either your pride could be hurt or you could gain recognition.

5._____You might feel that others are expecting too much from you or you could develop through interactions.

6._____Your career could be a focal point. Another person could take away a professional opportunity or offer you one.

7._____Others may anger you or you might be asked to take the initiative.

8._____Responsibility could be an issue in relationships. Someone could make you feel restricted, or another could provide security.

9._____You might feel that others are forcing you to talk or write too much, or you could enjoy communicating with others.

10._____You could find yourself evaluating the way you interact because of the way people are acting toward you.

11._____You might feel that some individuals are wasting your time or you could enjoy the pleasures of associating with others.

12._____ You could either have doubts about your own personal identity, or feel good about yourself because of the attitude of other people toward you.

13._____ You might think someone is trying to wield power over you or a relationship could cause you to analyze yourself or assert your power.

14._____ Strange people may come into your life or some individual might help you to assert your independence or utilize your creativity.

## Chapter Ten

# Putting It All Together

Thus far we have examined the transiting planets and nodes individually, but aspects from the transiting planets rarely occur in isolation. If you deal with transits in this manner, you could miss much that is in the total picture. You might concentrate on the motifs of one planet and not notice what else is activating your chart. For example, if transiting Saturn were squaring your natal Saturn, you could become so occupied with organizing, cutting back on activities, etc., that you might overlook the fact that Jupiter aspects were also in range. So you should pursue some form of growth or expansion as well. You need to balance the two principles in your life during that period. Once you form an overview, you can elaborate on the themes that are found and create a game plan for that time.

## Planetary Cycles and the Adult Stages

The cycles of the planets which can be combined with the psychological stages (mentioned earlier in this book) can provide a broad overview. For this you need only the birth year because that will tell you the age and, therefore, the phase in which the person is immersed. You will not know precisely what is occurring in the native's life, but you can use the tasks and types of shared experience given for each period in *The Seasons of a Man's Life* as a frame of reference. For example, the psychologists inform us that almost everyone in their late 30s and early 40s wants more freedom even if one had previously been totally satisfied. Astrologically, this correlated with transiting Uranus opposition natal Uranus, which makes the same statement at the same time. You cannot ascertain merely by the age alone in precisely which area the need for freedom will emerge in any given individual's life, but it will almost assuredly be an integral part of that age period.

Although the various phases were mentioned earlier, they were discussed mainly in terms of the planets singly. Below is a brief description of the psychological stages combined with the aspects of the Jupiter, Saturn, Uranus and Neptune cycles that occur during each period. This could form a backdrop or initial way to gain perspective on what generally is happening to someone before meeting with the person and examining the specifics. It is meant simply as a broad framework.

The first stage depicted in *The Seasons of a Man's Life* is the **Early Adult Transition**. This occurs between the ages of 17 and 22. The first task of this period is to leave the pre-adult world, which includes both ending relationships and also terminating situations that can be hindrances to establishing oneself in the adult world. This task can be connected with the transiting Saturn square natal Saturn which everyone experiences during this age span.

The second task is to take the initial steps necessary to move into the adult world and explore possibilities of how one will ultimately function there. This task is Uranian, and the first Uranus square Uranus also coincides with this period.

In other words, individuals between the ages of 17 and 22 could experience the freedom/responsibility dilemma. On the one hand, there is a need to sever old ties in order to be free, and on the other, there is a desire to establish oneself and become a responsible adult. Thus anyone in that age group has the potential of this dilemma. This by no means describes the entire life situation but can provide general information on which to build. You can easily check to see if the Saturn square Saturn and/or the Uranus square Uranus are in range at the time you are interpreting the transits. Often they coincide or overlap. If either aspect has occurred and will return, even if it is more than one degree away from exact, it could very possibly be relevant.

**Both** Saturn and Uranus are prominent for this age group, and a balance between the urges they represent is essential. Problems can arise if the individual is concentrating more on one than on the other—possibly clinging to something that should be dropped (Saturn) or else ignoring responsibilities for the sake of freedom (Uranus). If there is an imbalance, understanding the purpose of the period can be helpful, and, if necessary, the alchemical rituals associated with the planets can be used to restore equilibrium. If Saturn is the neglected planet, you can make lists to establish order, determine responsibilities and consider commitments. If Uranus is more dormant, you can find small ways in which to express freedom and be creative without rocking the boat too much. These interim measures can put you in touch with what should be done on a more meaningful level.

Each period of the adult cycle will contain at least one aspect of the Jupiter cycle and usually more. Jupiter is fairly regular in its movement and therefore will make a hard aspect to itself approximately every three years. The Early Adult Transition contains Jupiter opposition Jupiter (about age 18) and Jupiter square Jupiter (around age 21). The Jupiterian themes of development and personal growth form the underlying principles of the psychological stages, but specific tasks within each period are more easily associated with Saturn, Uranus and Neptune. Jupiter's most evident manifestation in the adult cycle is epitomized in the figure of the mentor, and it will be mentioned in that context. Even though Jupiter will not be discussed in detail within each phase, keep in mind that at least two of its cyclical aspects will be present each period, and use should be made of these aspects when they appear.

The stage called **Entering the Adult World** begins at about age 22 while one is still in the throes of Saturn square Saturn and Uranus square Uranus, both of which can continue until the age of 23. Therefore, the tasks described are the same as for the previous period, at least initially; but before the end of the period, around age 28, the shift from the pre-adult world to the adult world should be completed. The psychologists warn that if commitments are made before sufficiently exploring alternatives, an unsatisfactory life structure could result. But, conversely, if no commitments are made, no structure can be formed. This again suggests the necessity for balance between responsibility and freedom—Saturn and Uranus.

Other cyclic aspects which occur during this period are a Jupiter return around the age of 24 and Jupiter square Jupiter at about age 27. These aspects could be utilized to facilitate the move from the family of origin to one's own home base; or if the shift has occurred, it could be a time to expand the new structure in some way. The career might be developed or enhanced by a course of instruction. Knowledge could be acquired through travel. Then there could be the Jupiterian figure of the mentor in the picture to help the individual along. If the native feels overwhelmed during these Jupiter aspects, it should be pointed out that it is possible to select ways to grow for yourself, even if someone else is trying to direct you. By age 28, one's initial structure should have taken on form in terms of career, love relationships, friends, values and lifestyle. And possibly just in time for the **Age Thirty Transition**!

The **Age Thirty Transition** extends from about age 28 to 33. According to *The Seasons of a Man's Life,* this is a time "…to work on the flaws and limitations of the first adult life structure.…(People) modify their lives in certain respects, but they build directly upon the past and do not make fundamental changes. It is a time of reform not revolution" (p. 58).

Astrologically this period contains the Saturn Return (28-30). Jupiter opposition Jupiter (around age 30) and Jupiter square Jupiter (at about age 33). There are no cyclical Uranus aspects. Therefore, astrologically as well as psychologically, revolutionary tendencies would not be expected (unless shown by other aspects from transiting Uranus). This period is a bridge between the life structure of the first 30 years and that of the second 30 years. Thus it is the end of the beginning and start of the second period of the life. It can herald new commitments and responsibilities. But the changes are usually gradual and based on evaluation of past success. You might choose to make only minor changes and reaffirm your position or to tie up loose ends and move on. Even those who have never committed themselves prior to this time have a strong desire to do so. This transition seems to be easier for some than for others, and the alchemical rituals connected with Saturn and Jupiter should be kept in mind if needed,

**Settling Down** starts to take form at the end of the **Age Thirty Transition** (about age 33) and continues until around age 40. This is a period in which you want to establish yourself in society and attain success in the areas of your life that are of major importance to you. It could be in occupation, family, community, etc. There is a need for a feeling of internal success as well as acceptance and recognition in the world.

As you become more firmly entrenched in your position, the urge to be in total charge of your life becomes very strong, This usually occurs in the sub-period of **Becoming One's Own Man** ("Boom'}, which begins around age 36 and continues until the end of the **Settling Down** period at age 40. There is the desire to rid yourself of any mentors (the third Jupiter return at about age 36, and Jupiter square Jupiter at about age 39) who keep you from being the authority figure (Saturn square Saturn, ages 35-38). The need for freedom from former dependencies, such as detachment from the mentor, can be ascribed to the Uranus opposition Uranus which occurs between the ages of 39 and 43, as well as to the Jupiter aspects. The Uranus opposition

Uranus also plays an important role in the next period. In fact, it may be the initial indicator that a new transitional period is upon you.

The **Mid-life Transition**, according to *The Seasons of a Man's Life*, begins at around age 40 and lasts until about age 45. Astrologically, because of the cyclical planetary aspects we should probably expand this period to include age 39 on one end, and extend the other to age 46. Then this stage would totally encompass Uranus opposition Uranus (39-43), Neptune square Neptune (also possibly 39-43), and Saturn opposition Saturn (43-46). There are, of course, Jupiter aspects as well; but the tasks attached to this period can best be categorized under the headings of Uranus, Neptune and Saturn.

The **Mid-life Transition** is almost always associated with some form of change—sometimes drastic sometimes subtle—and this facet of the period can be ascribed to Uranus. There might be conspicuous changes such as a divorce, a new job, a change of residence. But even if there is no visible shift in the life, "If we look more closely…we discover seemingly minor changes…[have made] considerable difference. A man may still be married to the same woman, but the character of his familial relationships has changed appreciably for better or worse. Or the nature of his work necessarily changes in certain crucial respects during the Mid-life Transition" (p. 61).

It is very common during the Uranus opposition Uranus period to be restless, to want more freedom and to feel a strong urge to express your individuality. Some of the same feelings that were present during the **Early Adult Transition** and the early part of **Entering the Adult World** (which coincided with the Uranus square Uranus) are prominent here. However, in this period the desired changes may not be as easily handled as they were in the late teens and early twenties. At that time you are starting to create your own niche in life, and the ritual of leaving the family of origin is sanctioned by society. In fact it is an accepted part of moving into adulthood. But, by age 39 (the time of the onset of Uranus opposition Uranus), you are usually established in the world, and changes are frequently considered weaknesses. There is not often acceptance, let alone support, for someone who leaves a job or a spouse. Even the individual experiencing this aspect will sometimes wonder what great flaw in person in personality is making him or her so discontented. But here we are not only dealing with a different environment, we are also astrologically looking at a different aspect —an opposition instead of a square. With the square we should be coping with obstacles which can impede progress, but not necessarily looking back. With the opposition there is a need for balance of present assets with aspirations for the future.

Under the Uranus opposition Uranus, people often seem to be able to make changes that are long overdue. But there is another possibility. Because of the erraticism ascribed to Uranus, decisions could be made in haste, then regretted later. This is a time when the alchemical rituals connected to Uranus can be very valuable. They can help you keep your equilibrium while you are contemplating change. As stated in the Uranus chapter, it is often wiser to explore new avenues and be sure that the path ahead is clear before you close doors behind you.

This is also true of Neptune square Neptune which can occur at approximately the same time. Usually the Neptune square Neptune enters the scene a little later than the Uranus opposition Uranus, but with some birth years they appear simultaneously. Whether Neptune arrives at the same time as Uranus or closely follows it, the **Mid-life Transition** is obviously a period in which there can be instability—at least from the perspective of the mundane world. But one of the Neptunian tasks of this phase is to rise above the material—to get in touch with the spiritual side of your self. Any Saturn transits to the individual horoscope can be invoked to maintain balance as one experiences the nebulous qualities of Neptune.

Examination and blending of polarities within your self may be somewhat disconcerting but should be encouraged during this period. If you have ordinarily expressed the "masculine" or intellectual side of your nature, it could be difficult to incorporate the tender or demonstrative side that is more dormant within you. But once the process of internally integrating polarities has begun, you can feel more complete. This holds true of all the polarities mentioned in the Neptune chapter. Therefore, not only allow them to merge, but try to help them along. The alchemical rituals of Neptune that can be particularly helpful with this process are meditation and visualization.

The attachment/detachment polarity deserves special mention here because comprehension of it can help explain the dissatisfaction and feeling of failure that can be experienced during this period. Since there can be the desire to leave one's mark on the world, to have accomplished something that will live after you, material success and societal approval can seem less important than previously. The "de-illusioning" process is part of this syndrome. Material success may appear to have fallen short of expectations or may become irrelevant in one's total scheme of things. Let us stress that this is a time to "dissolve" factors that impede further development. Do not negate what has been.It is merely time to move on. Accepting this premise can assist you to progress rather than waste energy dwelling on the past.

After you have grappled with the Uranus opposition Uranus and the Neptune square Neptune, along comes Saturn opposition Saturn (ages 43-46). Ideally this indicates that after exploration and tentative changes have occurred (Uranus), and after undesirable qualities have been dissolved and new spiritual goals have been formed (Neptune), it is time to stabilize and evaluate your situation (Saturn). Whether or not you have dealt positively with the Uranian and Neptunian tasks, Saturnian matters will become prominent when the opposition is in range. Perhaps it is just time to place your life in order. But if you have not dealt effectively with the Uranus and Neptune transits, there will assuredly be dissatisfaction with the life structure. If this is the case, you can investigate the manner in which the Uranus opposition Uranus and Neptune square Neptune themes were expressed and determine what was left undone. Clues may be found in other aspects from transiting Uranus and Neptune at the time of the reading. For example, if transiting Neptune were aspecting natal Sun, there might still be unresolved ego issues that hinder spiritual development. Knowing that, you could reexamine that part of your character to determine action which could be taken to correct the situation.

Since Saturn represents the past, as the Saturn opposition Saturn forms, changes that occur will most likely be based upon past experience. Saturn indicates that one's present situation should be evaluated before moving on. Changes which may have coincided with the Uranus opposition Uranus and the Neptune square Neptune should be evaluated and either incorporated in the life or discarded. If the **Mid-life Transition** has been well-handled, the new structure is easily formed. If, however, the issues of freedom (Uranus) and spirituality (Neptune) have not found suitable expression, one can feel limited, and adjustments will be more difficult. The Saturn opposition Saturn is a time to improve and stabilize one's position.

This does not mean that you must wait for cyclical aspects of the outer planets to appear in order to make changes. What I have described are patterns frequently connected with particular adult stage. But changes can occur at any time. The movement of planets and their activation of our natal charts symbolize the continual opportunities that are offered to us to improve our positions. Sometimes we may take advantage of opportunities, sometimes we may act too rashly and sometimes we may not respond quickly enough. The greater our understanding of the choices available to us as expressed through the transits, the more likely that we will move in a desirable direction.

The stages described above take us through age 46. You might want to create subsequent phases using the symbolism of Jupiter, Saturn,

Uranus and Neptune as they are appropriate. One way this can be done is to look back to a similar aspect, which has already occurred, note its general implications at that time and apply it to the later period. If the aspects are not exact repetitions of each other, there will be some similarities, and some differences that have to be considered, as previously mentioned above with the Uranus square Uranus and the Uranus opposition Uranus.

For exact duplications such as the Saturn Returns, the first of which occurs between the ages of 28 and 30 and the second in the mid- to-late 50s, the similarities in attitude and experience are amazing in spite of the potential difference in the life position of two periods. With the second return people again start talking about what they are going to do when they grow up. In recent months a number of clients in the throes of their second Saturn Returns have informed me that they are taking early retirement in order to pursue a new occupation. Others are evaluating different sectors of their life such as marriage, children, home, etc., in much the same manner as they did during the first Saturn Return.

## Transiting Planets and Nodes in Houses

Another way to form an overview of the transits for a particular period is to place the transiting planets and nodes around the natal chart. When you see the houses in which the planets and nodes fall, you can determine to which areas you can apply the appropriate themes in terms of your life situation at that time. With this method you can use the inner as well as the outer planets. The inner planets will be indicative of immediate action, while the outer planets are associated with more long-term planning. As already stated, if you were to dwell on an outer planet transiting a particular house every day that it was in that area you could spend years concentrating on that one segment of your life and accomplish little because you would be ignoring the total picture.

The function of the **nodes** and that of each of the outer planets transiting houses have been given in their respective chapters. But the following is a synopsis with the addition of the inner planets. The time of a planet's transit through a house will, of course, vary according to the size of the house and possible retrogradation of the planet. The span given is based on a 30 degree house and is only a rough approximation. As a planet or node enters a house, you might want to activate the energies symbolized by the nodes, or that planet at least, in some small way.

The transiting **nodes** are in a house for approximately 1 2/3 years. Since the nodes form an axis, they will activate two houses simultaneously and inform you that relationships will be important in those particular oppositional areas during that time. As they move into houses, you might want to reach out to make contacts in one or both of these compartments. For example, if they are beginning to transit your second and eighth house, you could get in touch with someone from whom your income could be increased (second house), or offer a financial partnership to someone (eighth house), or both. If you choose not to form associations at the time of the nodes' entrance into houses, at least be sensitive to the possibility of opportunities that might come from others in terms of those areas during their stay in those houses.

Transiting **Pluto** will usually be in a house for at least twelve years. Keep in mind the concepts of power, analysis and transformation as it moves through each area. If some appropriate action can be taken as it enters a house, you could help the potential transformation along. For example, if you are planning to get married you could select the date of the wedding to coincide with Pluto's entrance into the seventh house. If nothing comes to mind, remember that the probable Plutonian message as it transits the seventh house is that the way you interact with other individuals may change over the next twelve or so years, or that certain intimate relationships could be transformed during that period of time. Analyze problems that arise with partners and be aware that power may be an issue in close associations. Above all, do not be passive nor allow resentments to build up with those who fall into the intimate category because that can lead to violent explosions.

Transiting **Neptune** is in a house for about 14 years. During that time you should become more highly evolved in terms of that area. As Neptune enters a house, you could make a conscious effort to do something spiritual. Artistic or altruistic, or rise above the material in some way in keeping with the symbolism of that compartment. If it is moving into the fifth house you might decide to release your influence over your children, especially if you have been domineering or overprotective. Or you could pursue some artistic project. If Neptunian motifs such as confusion and victimization surface in regard to the house Neptune is transiting, focus your attention on these matters. Then, either rise above them or try to dissolve them. In this way you can help to lift the Neptunian fog that tends to form in an area as Neptune transits that house in the horoscope.

Transiting **Uranus** will be in a house for about 7 years and there probably will be some obvious changes in that compartment during that time. Plutonian changes sometimes are subtle or internal, but Uranian changes are usually externally evident. As Uranus enters your sixth house, you might want to make adjustments in your daily routine activities, or add some form of creativity there. If you choose not to make some effort to change in that area, at least try to remain flexible there. Periodically you might find yourself interrupted or disrupted as you attempt to perform yours tasks. Or you could sometimes feel bored and change your routine without being urged from others. So whatever house Uranus is moving through, keep an open mind and be prepared to be as spontaneous as possible.

As **Saturn** moves around the chart, you have a choice of feeling frustrated and limited, or setting goals and working to attain them. Saturn will remain in each house for about 2-1/2 years and during that time you may experience the negative as well as the positive side of this planet, but organization, persistence and stamina **can** win out. If Saturn is entering your first house, you could become depressed and be strongly self-critical, but the message is really to improve yourself. Take stock of who you are, form a plan as to how you can make yourself better, and slowly, deliberately, take action. If you are overweight you could decide to go on a diet, but one that is slow and steady. A fad or crash diet will probably not work. Wherever Saturn is transiting, be as patient as possible. Progress may come at a snail's pace, but there definitely can be progress.

With **Jupiter** transiting a house you will have approximately a year to grow and develop in that area. As Jupiter is entering the second house, you might initiate some money-making scheme. You could also find during that one-year period that money comes in without much effort, but you should be wary of overextending yourself financially. No matter which house Jupiter is moving through you should try to expand your horizons, but try not to spread yourself too thin.

**Mars** transits a house in about two months—considerably less time than the planets discussed thus far. With the other outer planets it is easier to see the **total** results of a planet transiting a house after the fact. Periodically situations will arise that fit the definitions ascribed to both a planet and the house it is transiting, but when you are dealing with a time span of 1 to possibly 14 or more years, it is difficult to constantly categorize situations. In fact if you were to try to do this you might not have time to accomplish anything. With Mars, however, results are more immediate. As Mars enters a house, you can take the initiative or be assertive in terms of that area and see the effect of your actions shortly after it moves into the next area. For instance, as Mars enters your twelfth house you could delve actively into the subconscious and then, as it moves into the first, you could overtly make use of what you uncovered. As Mars transits a house, there will usually be noticeable activity in that area.

It is often difficult to separate the transits of **Venus, Mercury** and the **Sun** since they are never far from each other and can often be transiting the same house. They transit a house in about a month. Rarely will these planets indicate monumental happenings by themselves, but they can be incorporated in a grander scheme suggested by the outer planets. For example, if Saturn is transiting your tenth house, you might decide to work toward a promotion. You will have about 2-1/2 years to accomplish this task, and as Venus, Mercury and/or the Sun move into that area (which they do each year) you could plan appropriate activities to work toward your goal. With Venus you could socialize with your boss or use your charm. With Mercury you might express your brilliant ideas or communicate with influential people, and with the Sun find ways to make others aware of your presence and vitality.

The **Moon** travels around the chart in approximately 29 days and spends about 2-1/2 days in each house. Its movement is too rapid to warrant constant attention. Since the Moon represents emotionality and fluctuation, you could find that you are a little more sensitive in a particular area as the Moon transits that house. Or there could be some instability in matters pertaining to that house. It is also possible to consider the Moon as a trigger and you might select the time that it moves into a house to make changes in that area. As with Venus, Mercury and the Sun, it is most effective when coordinated with the outer planets.

## Customizing Transits

Thus far in this chapter transits have been discussed in their broadest terms. Now it is time to personalize them. The exact same transits may function quite differently for two individuals because their characters and personalities are different. Therefore, in order to make the best possible use of transits you must first have a basic understanding of the individual as depicted in the horoscope. You want an overview of how the native generally operates and how he or she can be effectively reached.

One way in which the *modus operandi* can be determined is to look for the signs, elements and modes emphasis in the chart. Sign emphasis can be ascertained by the combination of the following:

1. Sign of the Sun
2. Sign of the Moon
3. Sign on the Ascendant

4. Signs other than the Ascendant in the first house
      (because this is the house of the personality)
   5. Sign of the MC
   6. Stelliums by sign or house
   7. Sign emphasis by sign
   8. Sign emphasis by house

Numbers 1 through 5 are self-explanatory but 6, 7 and 8 might require some clarification. This can probably best be illustrated through an example. View the chart on the next page:

1. The Sun is in **Scorpio**

2. The Moon is in **Cancer**

3. The Ascendant is in **Aquarius**

4. Signs other than the Ascendant in the first house
   – all of **Pisces** and 14 degrees of **Aries**

5. MC in **Sagittarius**

6. A **stellium** is in **Scorpio** and in the **eighth house**. A stellium is 3 or more planets in the same sign or house. House stelliums reflect the sign that the house indicates in the natural zodiac and gives the underlying quality of that sign to the stellium. In this particular case the stellium is not only in the sign of Scorpio but also in its house, making Scorpio doubly strong. If the stellium had been in Scorpio in the seventh house, it would have counted as both a Scorpio stellium and a Libra stelllium and the Scorpio qualities would have been modified by Libra. If the stellium had contained one planet in Libra and two in Scorpio and been placed in the eighth house, it would not have counted as a Scorpio stellium by sign buy it would have by house. So it is possible to have a sign stellium, a house stelllium or both. The planets, by the way, need not be conjunct each other, just in the same sign or house.

7. **Sign Emphasis by Sign.** In this step the planets, the MC and the Ascendant are placed in their appropriate elements and modes, as can be seen above. Then the highest count by element and mode is noted. This will indicate a sign emphasis since no combination of element and mode are repeated. Here the highest element count is in water (five) and highest mode count is in fixed (five). Therefore, the sign of **Scorpio** is again emphasized.

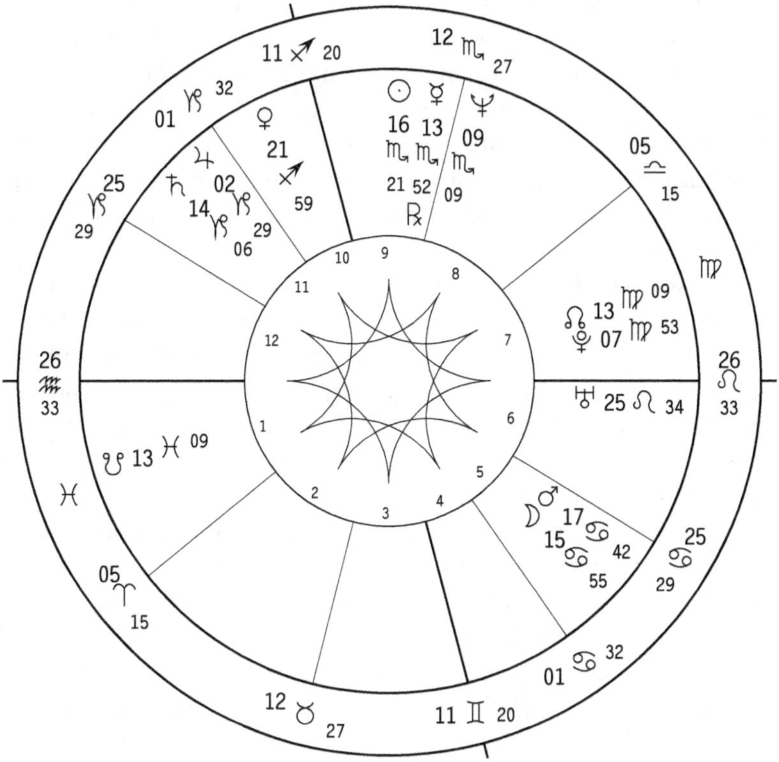

*Putting It All Together* 119

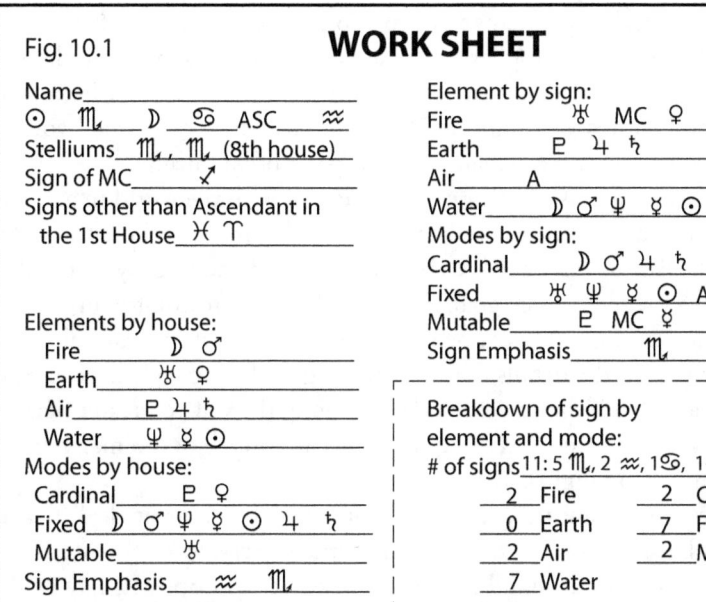

Fig. 10.1 **WORK SHEET**

Name_____
☉ ♏  ☽ ♋  ASC ♒
Stelliums ♏, ♏ (8th house)
Sign of MC ♐
Signs other than Ascendant in
 the 1st House ♓ ♈

Elements by house:
 Fire ☽ ♂
 Earth ♅ ♀
 Air ♇ ♃ ♄
 Water ♆ ☿ ☉
Modes by house:
 Cardinal ♇ ♀
 Fixed ☽ ♂ ♆ ☿ ☉ ♃ ♄
 Mutable ♅
Sign Emphasis ♒ ♏

Element by sign:
 Fire ♅ MC ♀
 Earth ♇ ♃ ♄
 Air A
 Water ☽ ♂ ♆ ☿ ☉
Modes by sign:
 Cardinal ☽ ♂ ♃ ♄
 Fixed ♅ ♆ ☿ ☉ A
 Mutable ♇ MC ☿
Sign Emphasis ♏

Breakdown of sign by
element and mode:
# of signs 11: 5 ♏, 2 ♒, 1 ♋, 1 ♐, 1 ♓, 1 ♈

 2 Fire      2 Cardinal
 0 Earth     7 Fixed
 2 Air       2 Mutable
 7 Water

8. **Sign Emphasis by House.** In this step planets are tallied in the houses in which they appear. The houses represent the signs with which they correlate in the natural zodiac. The first house represents Aries, the cardinal fire sign; the 2nd Taurus, the fixed earth sign; etc. The MC and Ascendant are not used because they are merely dividing lines when examining houses. In our example the highest counts are in the air and water elements (3 each) and in the fixed mode; so the signs emphasized here are **Aquarius** and **Scorpio**.

Next, count the total number of signs emphasized (11) and record how many times each sign is stressed, as shown on the worksheet, Fig. 10.1, page 119. Scorpio is emphasized 5 times (Sun, sign stellium, house stellium, emphasis by sign and emphasis by house; Aquarius twice (Ascendant and house emphasis); and Sagittarius (MC), Cancer (Moon), Pisces (sign other than the Ascendant in the first house) and Aries (sign other than the Ascendant in the first house) each once.

The final step in this method is to place the signs that are emphasized back into elements and modes. This gives you a concise profile of how the individual basically operates. With these 11 signs there are 2 fire signs emphasized (1 Sagittarius and 1 Aries), 0 earth, 2 air (2 Aquarius) and 7 water (5 Scorpio, 1 Cancer, 1 Pisces). Two are cardinal (1 Aries and 1 Cancer), 7 are fixed (5 Scorpio and 2 Aquarius) and 2 are mutable (1 Sagittarius and 1 Pisces).

If **fire** is the predominant element in your chart, you need to be actively involved. When problems arise, you don't usually sit back and wait to see what is going to happen. Rather, you take action, sometimes without thought of consequences (especially if Aries is the fire sign that is strong). Because of your zeal and enthusiasm you should be cautious not to overextend yourself and promise to deliver more than you are capable of producing.

**Earth** emphasis indicates that you deal with the world in a practical way. You want to understand and organize before you take action. You might make lists of what needs to be accomplished and step-by-step follow through to a predetermined goal. There needs to be a reason for any undertaking. The warning for earth people is that they should be careful not to get so involved in the details that they lose sight of the goal. This can most readily be avoided if they have small goals along the way to greater ones, so that results can be seen relatively quickly and there are not too many details in which to get entangled.

If you are an **air** person, you probably live a great deal in your head. You have the ability to create an abstract overview of situations and the objectivity to understand the positions of other people. It may be difficult, however, to take action (particularly for Librans and Aquarians) and so much time may be spent in forming a plan and talking about it that golden opportunities could be missed. Geminis, on the other hand, can be physically active, but typical Gemini activity usually does not produce meaningful results. Important actions still need to be planned and discussed before initiative is taken. Each of the air signs has a different manner in which they can be stimulated into action—Geminis through having a variety of tasks to perform that they can alternate; Librans by letting them know they are helping someone close to them. (Libra is after all a cardinal sign); and Aquarians by correlating action with a goal, particularly a humanitarian one.

**Water** people are intuitive and tend to operate on "gut feeling." If something "feels" right to a water person action can be taken. These individuals do not need to think things through nor make sense of situations, and often you will hear them make such statements as, "I just know it in my heart!" The caution here is not to let your feelings totally take over because they tend to color a situation, and there can be a discrepancy between objective reality and your perception of it. Fortunately, no individual is composed of only one element, and the other prominent elements can be called upon if such difficulties arise.

If the **cardinal** signs are strong in a chart, this is one indication that you are an initiator. As with the fire signs, cardinal signs want to be actively involved. One need only find the right approach. The reasons for taking the initiative among the four cardinal signs are quite different. Aries take the initiative for self. Capricorn is stimulated by the thought of status or professional advancement. Neither Cancer nor Libra seems as aggressive as the other two cardinal signs, but, as stated above, Librans can take action for those they are close to. And Cancerians need no prodding when protecting home or family.

The **fixed** signs are tenacious and persistent. Those with the fixed signs emphasized have the ability to remain with people and situations for long periods of time much longer than with either of the other modes. They can be creatures of habit, and therefore, extremely resistant to change. Even Aquarius, the supposed revolutionary sign, wants to make sure that the end result will make the war worthwhile. The customary response to any suggested alteration is initially "no." Those who have Aquarius emphasized in their charts may move a little more rapidly than Taureans, Leos and Scorpios, but none of the fixed signs look to change with eager anticipation. It is important for these individuals to know that you need not move drastically and rashly. And this group especially finds the alchemical rituals appealing.

Those who have the **mutable** signs strong in their charts are flexible and, at least theoretically, seem to welcome the idea of change. Whatever signs, elements and modes are emphasized, all human beings are frequently apprehensive of totally new conditions. Fortunately, however, immediate, sweeping changes are not usually necessary. Because of the inconsistency between eager anticipation of change and somewhat innate resistance to it, mutable individuals are capable of false starts. Therefore it might take some time to move definitively, but they do have the advantage of being able to shift direction. The secret is to keep them from going around in circles. Here again the appropriate alchemical rituals can help them to focus.

There is no such thing as a birth chart that consists of all planets and points in a single sign, element and mode. (At least I have never seen one.) Each horoscope is composed of various combinations, and this should be taken into consideration as we delineate natal charts. In our above example, the strongest element is water and the strongest mode is fixed. Thus this individual most likely operates intuitively (water), and once he has embarked on a path, it is probably difficult to get him to change direction (fixed). Even though this is his customary way of taking action, it is possible for him to be somewhat objective and to do some planning (air). And with the fire he would move a little more quickly than someone who does not have fire stressed at all. There is also enough cardinality to take the initiative more rapidly than the fixity would indicate, and enough mutability to avoid total inflexibility.

The particular signs that are emphasized in the horoscope shed further light on the behavior of the native. Scorpio is the strongest sign, and therefore characteristics ascribed to it will be the most noticeably manifested. So, along with operating on a "gut level," this individual probably analyzes situations deeply, keeps a great deal inside, etc. Aquarius as the second strongest sign will add that he will probably attempt to look objectively at the information he has acquired and put it together mentally, in his own way, probably without consultation. Next you would include the qualities attributed to the other signs that are emphasized in the chart, and, thus, form a cohesive picture of the *modus operandi* of this person.

Since the definition of transits is the dynamics of planets through the zodiac, it is logical that they are associated with action (either internal or external) in our lives. Therefore, examination of Mars should be added to the element, mode and sign interpretation. Mars by sign, house and aspect will provide additional information as to how the individual will initiate and perform. In our example Mars reinforces the idea that the native is motivated by feelings and intuition. Mars is not only in Cancer, but it is also conjunct the Moon in Cancer. The opposition of Saturn in Capricorn (its own earth sign) to these two planets might compensate for the low earth count in the chart. It would modify the emotionality and indicates that even though his reaction to situations could be highly emotional, he would need to balance his feelings with practical reasons for taking action. He could probably most easily initiate for others (Mars in Cancer and sextile north node in the seventh house) but he will undoubtedly think of his own gratification as well (Mars trine Sun). The trine of Mars to Mercury could indicate that he would talk more readily about what he was going to do than the Scorpio emphasis would imply.

This is by no means a total interpretation. Nor is it the only way to start a delineation. The purpose of the illustration is simply to show one method, which can be employed to explain how someone can most easily make use of transits. We are different from each other, and this must be considered when trying to make the most of transits. You would not, for instance, tell people who are impatient to take their time. Instead you would help them find ways to complete tasks quickly.

You should also look for important messages or themes within the natal chart, with particular attention paid to potential dilemmas. You might, for example, have a T-square, which is activated by a transiting planet. This could mean that you were being offered an opportunity to face and work through a possibly difficult life pattern. Or perhaps there is a repeated motif such as a conjunction of Saturn and Uranus compounded by Sun in Capricorn and Moon in Aquarius—two statements of a possible freedom-responsibility dilemma. And let us say that transiting Saturn is conjoining the Saturn-Uranus conjunction. It would, of course, be Saturn Return time, but the freedom-responsibility dilemma could be activated as well. Seemingly conflicting situations might arise. Knowing the purpose of the issues and recognizing that both polarities need to be accommodated can help you deal more quickly and rationally with circumstances that fit the symbolism.

Once the natal chart has been interpreted, there are a number of ways in which investigation of the transits can be organized. Whatever method is employed, a preliminary step should be taken. Place the natal planets and points in numerical order, regardless of sign. This will simplify the task of following the transits.

For our example, see the list at the top of the next page:

|        |   |             |     |   |            |
|--------|---|-------------|-----|---|------------|
| ♃      | — | 02 ♑ 29     | ☽   | — | 15 ♋ 55    |
| ♇      | — | 07 ♍ 53     | ☉   | — | 16 ♏ 21    |
| ♆      | — | 09 ♏ 09     | ♂   | — | 17 ♋ 42    |
| MC     | — | 11 ♐ 20     | ♀   | — | 21 ♐ 59    |
| ☊      | — | 12 ♍ 10     | ♅   | — | 25 ♌ 34    |
| ☿      | — | 13 ♏ 52 ℞   | ☽   | — | 26 ♎ 07    |
| ♄      | — | 14 ♑ 06     | ASC | — | 26 ♒ 33    |

You can also add the asteroids, Chiron and any other points that you consider important. Then, looking at a current ephemeris, you can quickly see which transiting planets are forming exact aspects to natal planets and points. With our example chart, if you were to see in a particular month that transiting Saturn were moving between 21 and 25 degrees of Aquarius, you would know at a glance that it was sextiling natal Venus, opposing natal Uranus, and trining natal Part of Fortune. But some form of transit sheet can offer a more cohesive representation.

## Monthly Transit Sheets

You can create a monthly transit sheet with a piece of 8-1/2 by 11 inch lined paper (preferably with narrow spaces). At the top of the page write the month and year and down the left-hand side the dates and corresponding days of the week. Then, place the planets and points in numerical order on the right-hand side of the page as shown in Fig. 10.2 on the next page.

With the monthly transit sheet it is most convenient to use an ephemeris that is calculated for the local time zone. The reason for this is that we are using aspects from all the transiting planets (except the Moon), and since the inner planets ordinarily move quickly, it is possible to be a day off with these aspects if an ephemeris set for Greenwich is used. The monthly transit sheet is an excellent tool for the classroom. Its main purpose is to develop an understanding of how the definitions ascribed to planets and points work in transit and in the individual horoscope.

In Fig. 10.2 the first planet in each set is the transiting planet, the other planet or point in the pair is from the natal chart, and the aspect formed is placed between them. Because this is a teaching tool for beginners, only the 30 degree aspects are used as they can easily be seen. This does not, however, negate the importance of the less obvious aspects such as the semisquare and sesquiquadrate, and you might want to add them. The aspect is drawn on the date on which it is exact. With the aspects from the outer planets and nodes, the line above the aspect starts on the date on

## September 1986

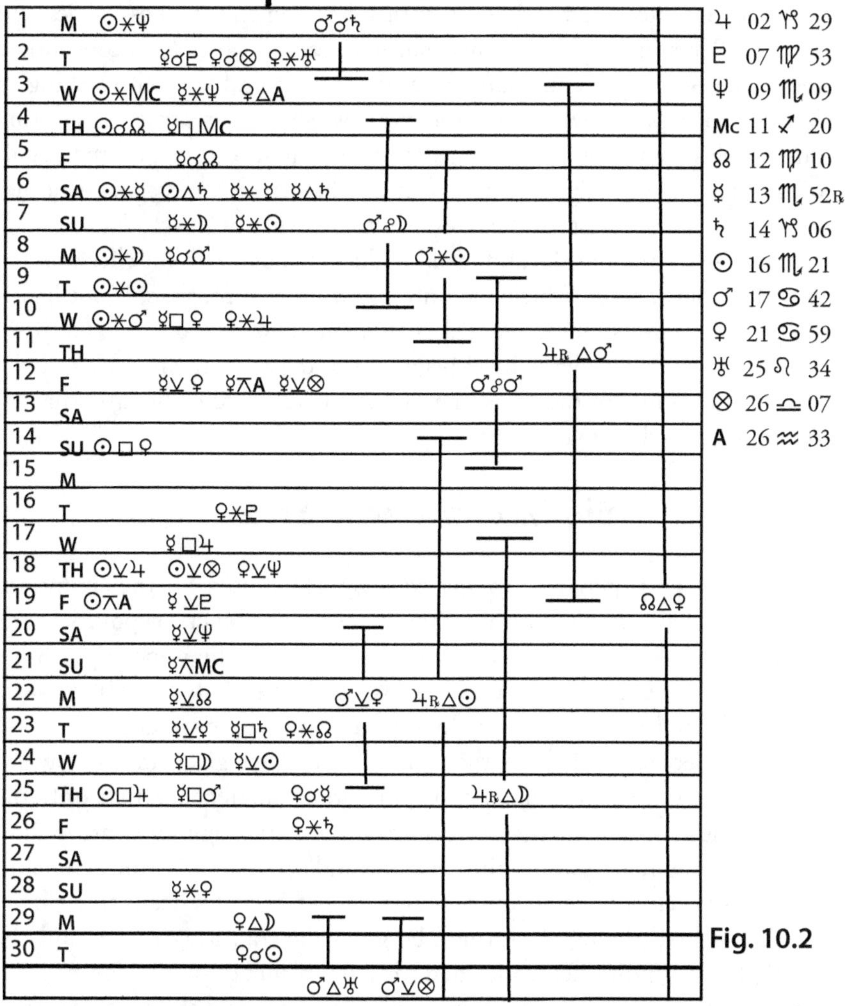
1 M ☉⚹♆ ♂☌♄
2 T ☿☌♇ ♀☌⊗ ♀⚹♅
3 W ☉⚹MC ☿⚹♆ ♀△A
4 TH ☉☌☊ ☿□MC
5 F ☿☌☊
6 SA ☉⚹☿ ☉△♄ ☿⚹☿ ☿△♄
7 SU ☿⚹☽ ☿⚹☉ ♂☍☽
8 M ☉⚹☽ ☿☌♂ ♂⚹☉
9 T ☉⚹☉
10 W ☉⚹♂ ☿□♀ ♀⚹♃
11 TH ♃℞△♂
12 F ☿⊻♀ ☿⊼A ☿⊻⊗ ♂☍♂
13 SA
14 SU ☉□♀
15 M
16 T ♀⚹♇
17 W ☿□♃
18 TH ☉⊻♃ ☉⊻⊗ ♀⊻♆
19 F ☉⊼A ☿⊻♇ ☊△♀
20 SA ☿⊻♆
21 SU ☿⊼MC
22 M ☿⊻☊ ♂⊻♀ ♃℞△☉
23 T ☿⊻☿ ☿□♄ ♀⚹☊
24 W ☿□☽ ☿⊻☉
25 TH ☉□♃ ☿□♂ ♀☌☿ ♃℞△☽
26 F ♀⚹♄
27 SA
28 SU ☿⚹♀
29 M ♀△☽
30 T ♀☌☉
♂△♅ ♂⊻⊗
♃ 02 ♑ 29
♇ 07 ♍ 53
♆ 09 ♏ 09
Mc 11 ♐ 20
☊ 12 ♍ 10
☿ 13 ♏ 52℞
♄ 14 ♑ 06
☉ 16 ♏ 21
♂ 17 ♋ 42
♀ 21 ♋ 59
♅ 25 ♌ 34
⊗ 26 ♎ 07
A 26 ♒ 33

**Fig. 10.2**

which the transiting planet or node moves to within 1 degree of the natal planet or point. The line below the aspect ends when the transiting planet or node moves more that 1 degree away from the natal planet or point.

As you begin to draw monthly transit sheets, you might want to keep an accompanying diary or at least note anything that occurs which fits the planetary symbolism for the appropriate time. This can offer clues as to how you can make the most of the messages provided by the transits. For example, one student with Venus in Virgo in the tenth house found, through a few months observation, that each time Venus was aspected by a transiting planet something significant happened in his work. Rarely were his wife, his artistic ability or sociability involved. Thereafter, when he saw such aspects, he made professional contacts at that time rather than focusing on other possibilities.

When using the monthly transit sheets it is possible to concentrate too much on daily events. Sometimes you might find yourself reaching to find an appropriate event on the day on which, let us say, the transiting Sun sextiles natal Venus; and you become so intent on making the details of the transits work that you lose sight of the broader messages—in other words, losing the forest for the trees. Or there could be the tendency to rely too heavily on the transits and avoid making a move without checking the monthly transit sheet, which certainly could impede progress, if not halt it altogether. Transits should be used as guidelines, not as prophecy. For these reasons, after the students have begun to understand themselves and the symbolic meaning of the transiting planets, I suggest they use a six-month transit sheet as their main tool.

## Six-Month Transit Sheets

The same type paper (narrow-spaced 8 by 11) is used for the six-month transit sheet as for the monthly transit sheet. The planets or points of the natal chart are again placed in numerical order on the upper right-hand side of the page. Across the top, however, is written only the year or years for which the transits are being done instead of the month and year. And down the left-hand side are the months and dates given in five-day intervals instead of daily (Fig. 10.3). One other difference is that only the Moon's mean nodes and the planets from Jupiter through Pluto are used—the transiting planets and points that are indicative of milestone developments. The setup of the transits is also similar, with the transiting planet appearing first, the aspect formed between the transiting planet and the natal planet or point next, and then the natal planet or point. The aspect is drawn during the 5-day period in which it is exact, and the 1 degree range of the aspect is still shown by the lines above and below the aspect. In our example, where transiting planets go direct, and are in range of an aspect, the letter "D" is written. Had any planets been going retrograde, an "R" would have been placed where they went retrograde. When numbers appear near the "D" or "R" this means that the aspect is within 1 degree but will not be exact. For example, the 33' written above the Neptune conjunct Jupiter means that Neptune moves to thirty-three minutes of exact before it goes direct.

The main purpose of the six-month transit sheet is to obtain an overview for the period so that you can prepare and make the most of the

*Putting It All Together* 127

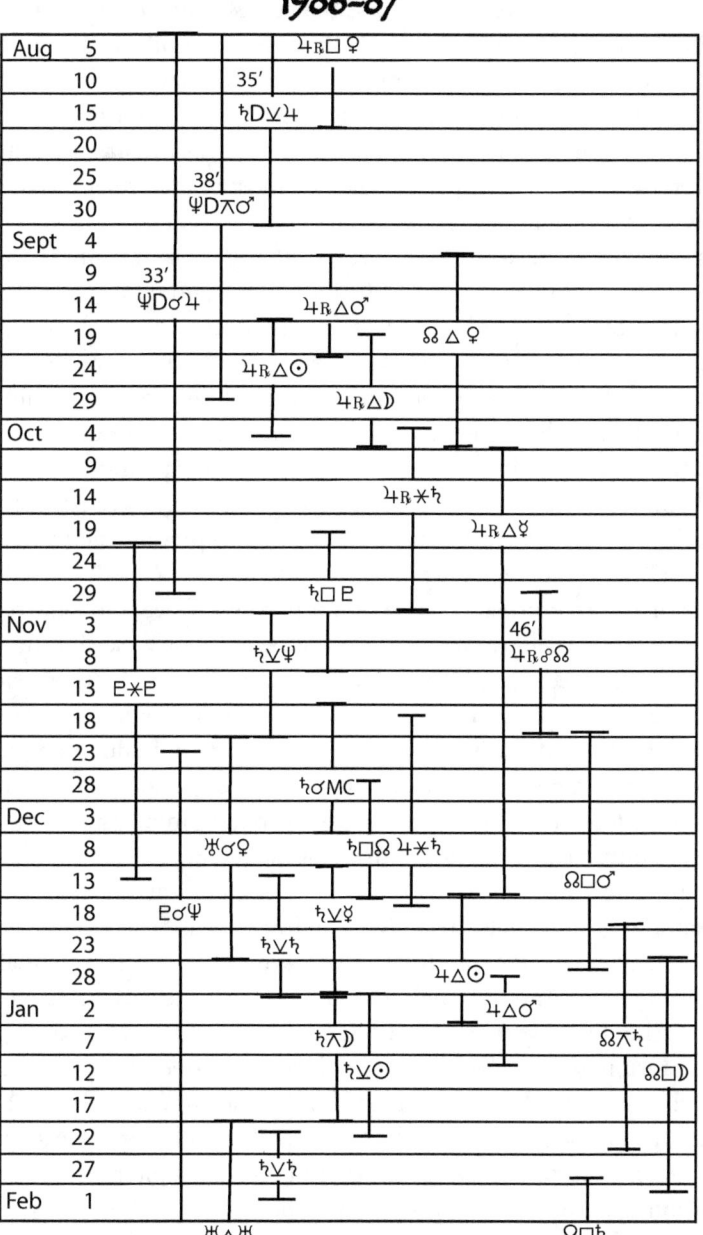

**Fig. 10.3**

themes presented. Why a six-month transit sheet as opposed to a year or two or ten? Because what you do with this six-month period will affect the next six months. Your situation then might be quite different from the way it is now. And whether or not you are satisfied with your circumstances will depend upon your application and response to the transits. In other words, a six-month transit sheet can help you make the most of the present so that the future will be more gratifying.

As you interpret, keep the native's *modus operandi* in mind. You can also note the psychological stage of the individual, and if he or she is of an age at which any of the hard aspects of the outer planetary cycles could be formed. In our example the native was born in 1960. Therefore, he is in **Entering the Adult World**—a period in which he should be establishing himself in the adult world. At age 25 to 26 he is past the Saturn square Saturn and Uranus square Uranus. The unsettled part of the period (Uranus) should be behind him, and his goals for this stage (Saturn) have probably been formed. We now have our background for interpreting the transits and creating a "game plan" for the period.

The first step in utilizing the six-month transit sheet is to view the overall picture without delineating the specific aspects. Look to see if certain months have more aspects than other months. Does a group of aspects end and another begin at a particular time? If this is so, that would be a good point at which to shift emphasis or move in a new direction. In our example, October seems to be a month in which some aspects end and others begin; and November, December and January seem to contain more aspects than the previous months. This could mean that August and September could be preparation time and, perhaps, October could be launch time for a plan.

Examining the aspects themselves will further clarify and define the situation. Remember that the transiting planets symbolize certain potentials that are available to us. And the natal planets and points they activate advise us of what we should be working on within ourselves. If a particular transiting planet is forming a number of aspects during a period, the principles ascribed to that planet will be prominent during that time and should be applied. In our example, in the months of November and December there are a number of aspects formed by Saturn and also by Jupiter. Therefore, both expansion and contraction will probably be factors in those months. And this person should know that a balancing act will be effective at that time—exposing himself to situations that will help him to develop (Jupiter) and then clarifying and organizing (Saturn) what he is learning.

As we look at the December and early January period, we see that Saturn and Jupiter are both activating his natal Sun and Moon, as are the transiting nodes. This could indicate that whatever he is exposed to during that time could have an impact on his ego and/or emotions. With the transiting nodes being involved, others may focus on the facets of character represented by the Sun or Moon. Being armed with all this information he could determine ways in which he would like to develop and solidify his feelings (Moon) and attain ego gratification (Sun). Then, when the aspects are in range, he could take appropriate action.

In analyzing the six-month transit sheet you could also incorporate the houses in which natal and transiting planets and points are posited and the houses they rule. And the more information you have about the present life circumstances of the native, the more specific you can be in the interpretation of the transits. This data, however, should not be used to predict the future, but rather to create it.

## Alternatives

Two alternatives to the six-month transit sheet are the 30 degree and 45 degree graphic ephemerides. These show graphically the movement of the nodes and all the planets but the Moon for the period of one year. Also indicated are solar and lunar eclipses and retrogradation including stations of the planets. On the 30 degree ephemeris (See Figure 10.4 on the next page), you will see that the numbers 0, 5, 10, up to 30 appear on both sides. With your planets in numerical order it is simple to set up the page for your own chart. You merely take a ruler and draw a line at the appropriate place. Write the glyph for that natal planet or point next to the line. As you then look across, you can see all of the 30 degree aspects that planet or point makes with the transiting planets and nodes in that year. As you place all of your planets and points on the sheet, you can also look from top to bottom on a particular date and see what is occurring on that day or during a particular segment of the year.

---

(Ed. Note: In the first printing of this book, the author recommended pads of graphic ephemerides that are no longer available, so far as we can discover. Most current full service home software programs include graphic ephemerides, and of course, they are also available from Astro Computing Services. The author intended (and we agree) that instruction on this valuable technique is best served by seeing the graphics both with and without the natal lines—the better to learn by doing. Among our available options, *IO Edition* for Macintosh, Time Cycles Research, has half-year graphic ephemerides that offer the greatest clarity for book production, so we have run our illustrations of the 30° and 45° graphs without natal lines from IO. Our thanks to IO's developer, Dennis Haskell, *www.timecycles.com* .

130  *Astrological Alchemy*

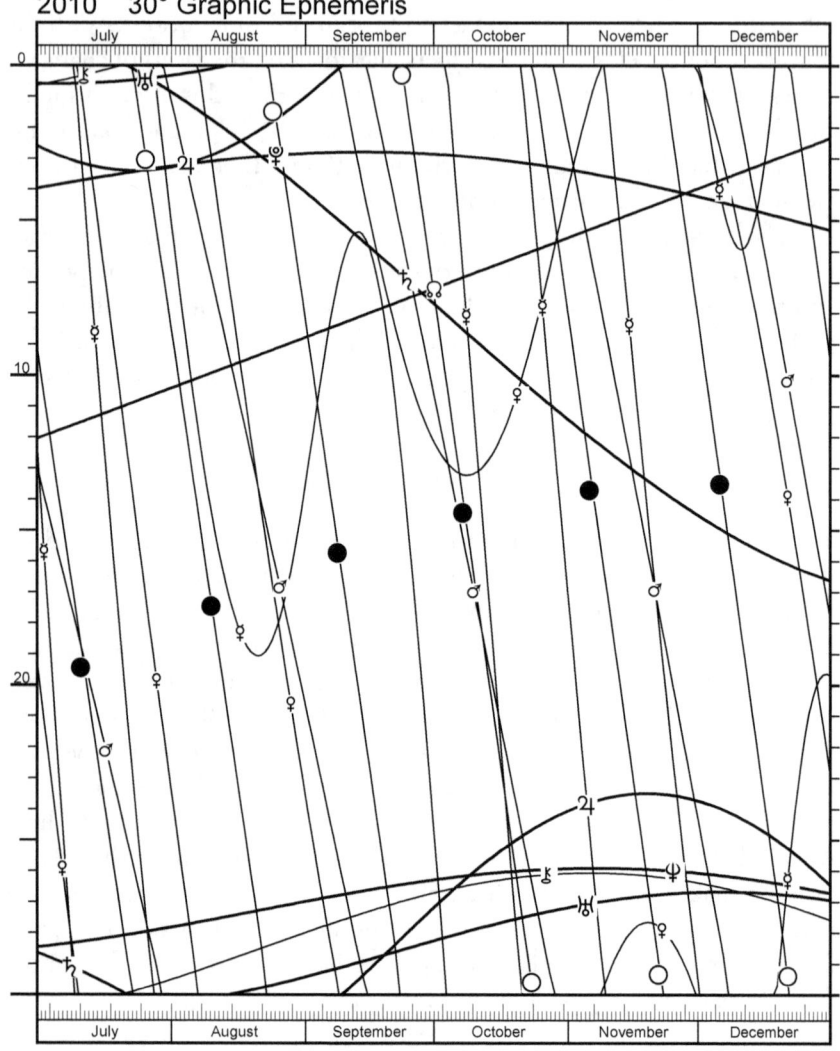

**Fig. 10.4**

The 45 degree ephemeris (Fig. 10.5 on the next page), is set up to show only the hard aspects which include the conjunction, opposition, square, semisquare and sesquiquadrate. It is slightly more complex to set up than the 30 degree ephemeris. Here too, you must add the natal planets in numerical order but also according to mode. It is clearly marked where each mode begins so that the cardinal section begins at the first 0 degree (note the 0C for Cardinal) and continues with 0M for Mutable, then 0F

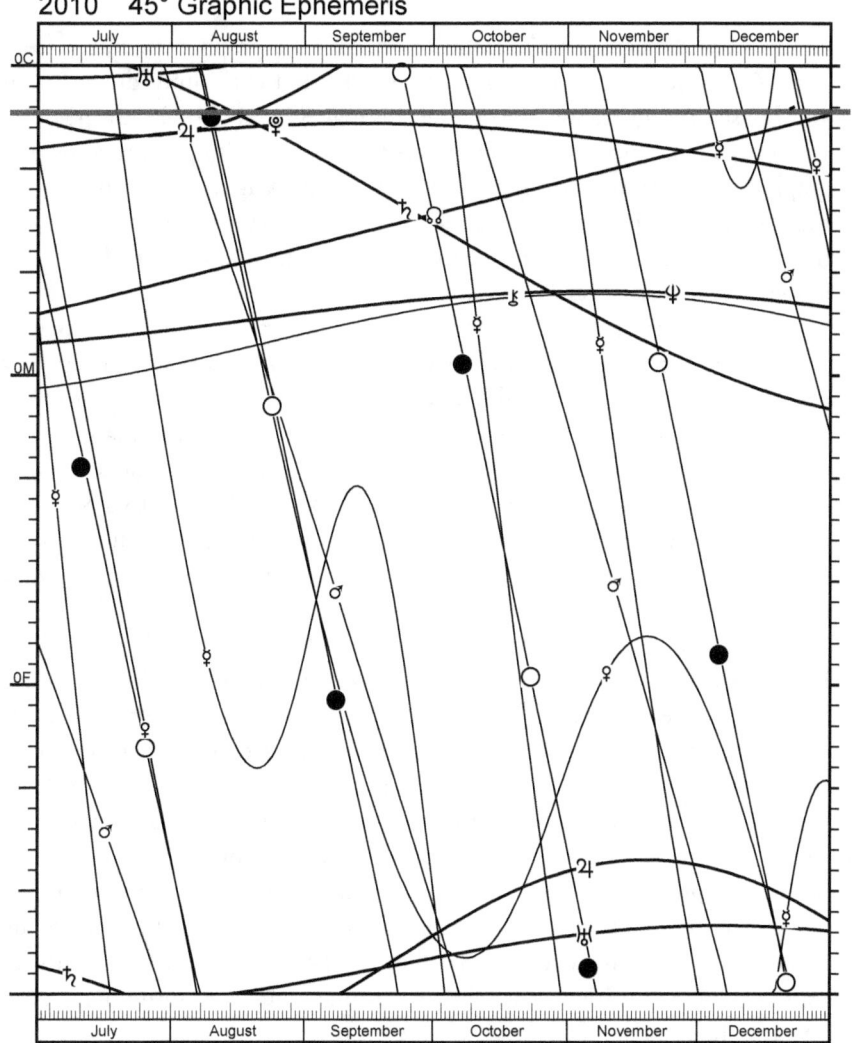

**Fig. 10.5**

for Fixed. Lines dividing each section into thirds represent 10° and 20° and each small line marks 2°. Since Jupiter is 2 Capricorn 29, its line would be drawn slightly beneath the first marker (2 degrees) of the cardinal signs (as shown by the grey horizontal line—and we can then see that in early August natal Jupiter is aspected by an eclipse and then by transiting Jupiter and Pluto. If you prefer to have the work of drawing in natal planet lines done for you, there are plenty of options either through Astro Computing Services or from various of the higher end astrological software programs currently on the market.

On the next page is an illustration of a 45° graphic ephemeris from Astro Computing Services that includes the natal chart lines drawn in, as you can see, straight across the graph. This makes it easy to scan for the times when each natal placement is intersected with a transiting planet line. The graphic ephemeris is reproduced here in black and white, and is considerably reduced in size. The actual product is printed in color, for extra clarity in seeing the difference between the transiting planet lines and the natal lines. This graphic ephemeris covers one full year.

Another interesting manner in which transits can be displayed is the monthly transit calendar, which could replace the monthly transit sheet. This calendar includes the lunar aspects to your natal chart with time of exactness, as well as the aspects from the other transiting planets. The transiting aspects are simply listed each day in a typical calendar format. An example of this option, also available from Astro Computing Services, is show on page 134. (Fig. 10.7). Note that the entire page is not shown. The last week has been cut off, so as to keep the size of the illustration large enough to easily read the aspect listings.

As you can see, there are a number of ways in which transits can be viewed. There are also choices as to how you can use the material in this book. It can serve as a guide in teaching a transit class. The ten chapters could provide the material for a ten-week course. The quizzes at the end of the chapters can be used as homework assignments. They can help students learn more about transits and also provide a way to stimulate class discussions. Personally, you might consider the book a source for broadening your concept of transits and adding to your repertoire. Or use the information to better understand what is happening to others. But, above all, **use** transits. Learn for yourself that they need never be limiting—that they represent opportunities by which we can make our lives more productive and fulfilling.

Answers to Questions 133

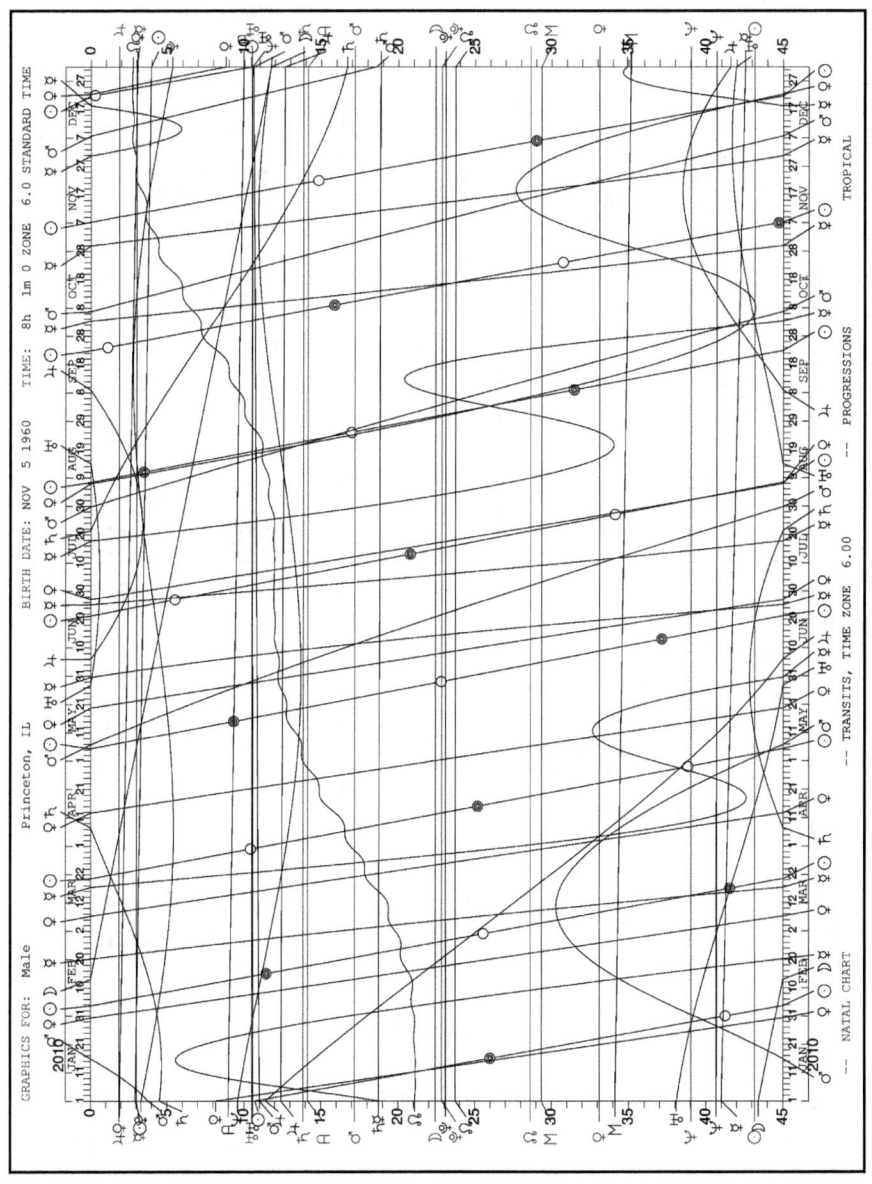

**Fig. 10.6**

45° Graphic Ephemeris with Natal Chart Lines

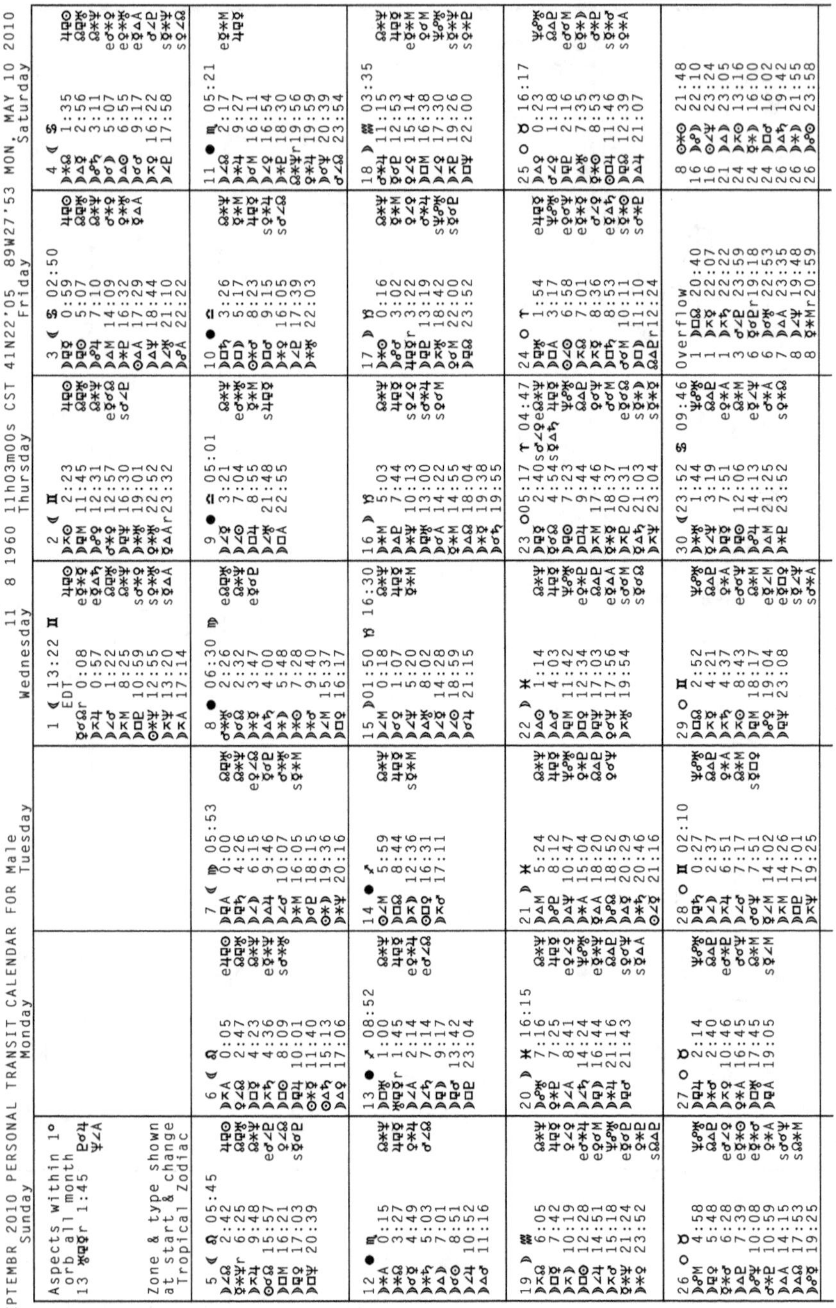

**Fig. 10.7**

*Portion of Calendar Transit Page for September 2010*

# Chapter Eleven

# Answers To Exercises

## Answers to Mars Exercises

### Mars transiting houses
1. d. Mars in the 4th house
2. i. Mars in the 9th house
3. l. Mars in the 12th house
4. b. Mars in the 2nd house
5. e. Mars in the 5th house
6. a. Mars in the 1st house
7. j. Mars in the 10th house
8. g. Mars in the 7th house
9. c. Mars in the 3rd house
10. f. Mars in the 6th house
11. k. Mars in the 11th house
12. h. Mars in the 8th house

### Mars in aspect
1. l. Mars-Ascendant
2. a. Mars-Sun
3. b. Mars-Moon
4. h. Mars-Uranus
5. n. Mars-nodes
6. m. Mars-Part of Fortune
7. c. Mars-Mercury
8. k. Mars-MC
9. f. Mars-Jupiter
10. e. Mars-Mars
11. i. Mars-Neptune
12. d. Mars-Venus
13. j. Mars-Pluto
14. g. Mars-Saturn

## Answers to Jupiter Exercises

### Jupiter transiting houses
1. i. Jupiter in the 9th house
2. g. Jupiter in the 7th house
3. e. Jupiter in the 5th house
4. a. Jupiter in the 1st house
5. j. Jupiter in the 10th house
6. h. Jupiter in the 8th house
7. k. Jupiter in the 11th house
8. b. Jupiter in the 2nd house
9. d. Jupiter in the 4th house
10. l. Jupiter in the 12th house
11. c. Jupiter in the 3rd house
12. f. Jupiter in the 6th house

### Jupiter in aspect
1. n. Jupiter-nodes
2. e. Jupiter-Mars
3. f. Jupiter-Jupiter
4. d. Jupiter-Venus
5. g. Jupiter-Saturn
6. l. Jupiter-Ascendant
7. c. Jupiter-Mercury
8. h. Jupiter-Uranus
9. a. Jupiter-Sun
10. i. Jupiter-Neptune
11. b. Jupiter-Moon
12. k. Jupiter-MC
13. j. Jupiter-Pluto
14. m. Jupiter-Part of Fortune

## Answers to Saturn Exercises

### Saturn transiting houses
1. f. Saturn in the 6th house
2. i. Saturn in the 9th house
3. b. Saturn in the 2nd house
4. c. Saturn in the 3rd house
5. e. Saturn in the 5th house
6. d. Saturn in the 4th house
7. h. Saturn in the 8th house
8. a. Saturn in the 1st house
9. j. Saturn in the 10th house
10. l. Saturn in the 12th house
11. g. Saturn in the 7th house
12. k. Saturn in the 11th house

### Saturn in aspect
1. g. Saturn-Saturn
2. n. Saturn-nodes
3. b. Saturn-Moon
4. m. Saturn-Part of Fortune
5. l. Saturn-Ascendant
6. a. Saturn-Sun
7. d. Saturn-Venus
8. h. Saturn-Uranus
9. k. Saturn-MC
10. c. Saturn-Mercury
11. e. Saturn-Mars
12. i. Saturn-Neptune
13. j. Saturn-Pluto
14. f. Saturn-Jupiter

# Answers to Uranus Exercises

## Uranus transiting houses
1. i. Uranus in the 9th house
2. e. Uranus in the 5th house
3. c. Uranus in the 3rd house
4. j. Uranus in the 10th house
5. k. Uranus in the 11th house
6. d. Uranus in the 4th house
7. a. Uranus in the 1st house
8. f. Uranus in the 6th house
9. g. Uranus in the 7th house
10. b. Uranus in the 2nd house
11. l. Uranus in the 12th house
12. h. Uranus in the 8th house

## Uranus in aspect
1. c. Uranus-Mercury
2. k. Uranus-MC
3. g. Uranus-Saturn
4. l. Uranus-Ascendant
5. h. Uranus-Uranus
6. n. Uranus-nodes
7. e. Uranus-Mars
8. j. Uranus-Pluto
9. m. Uranus-Part of Fortune
10. f. Uranus-Jupiter
11. i. Uranus-Neptune
12. b. Uranus-Moon
13. a. Uranus-Sun
14. d. Uranus-Venus

# Answers to Neptune Exercises

## Neptune transiting houses
1. g. Neptune in the 7th house
2. b. Neptune in the 2nd house
3. a. Neptune in the 1st house
4. f. Neptune in the 6th house
5. e. Neptune in the 5th house
6. i. Neptune in the 9th house
7. h. Neptune in the 8th house
8. c. Neptune in the 3rd house
9. j. Neptune in the 10th house
10. l. Neptune in the 12th house
11. k. Neptune in the 11th house
12. d. Neptune in the 4th house

## Neptune in aspect
1. c. Neptune-Mercury
2. n. Neptune-nodes
3. d. Neptune-Venus
4. a. Neptune-Sun
5. l. Neptune-Ascendant
6. m. Neptune-Part of Fortune
7. j. Neptune-Pluto
8. g. Neptune-Saturn
9. k. Neptune-MC
10. b. Neptune-Moon
11. e. Neptune-Mars
12. i. Neptune-Neptune
13. f. Neptune-Jupiter
14. h. Neptune-Uranus

# Answers to Pluto Exercises

## Pluto transiting houses
1. j. Pluto in the 10th house
2. f. Pluto in the 6th house
3. c. Pluto in the 3rd house
4. k. Pluto in the 11th house
5. a. Pluto in the 1st house
6. g. Pluto in the 7th house
7. l. Pluto in the 12th house
8. b. Pluto in the 2nd house
9. i. Pluto in the 9th house
10. d. Pluto in the 4th house
11. h. Pluto in the 8th house
12. e. Pluto in the 5th house

## Pluto in aspect
1. c. Pluto-Mercury
2. k. Pluto-MC
3. e. Pluto-Mars
4. n. Pluto-nodes
5. i. Pluto-Neptune
6. a. Pluto-Sun
7. h. Pluto-Uranus
8. l. Pluto-Ascendant
9. b. Pluto-Moon
10. g. Pluto-Saturn
11. m. Pluto-Part of Fortune
12. j. Pluto-Pluto
13. f. Pluto-Jupiter
14. d. Pluto-Venus

# Answers to Node Exercises

## Node Transiting Houses
1. b. node in the 2nd house
2. j. node in the 10th house
3. g. node in the 7th house
4. f. node in the 6th house
5. e. node in the 5th house
6. k. node in the 11th house
7. c. node in the 3rd house
8. l. node in the 12th house
9. i. node in the 9th house
10. a. node in the 1st house
11. h. node in the 8th house
12. d. node in the 4th house

## Node in Aspect
1. b. node-Moon
2. i. node-Neptune
3. m. node-Part of Fortune
4. a. node-Sun
5. f. node-Jupiter
6. k. node-MC
7. e. node-Mars
8. g. node-Saturn
9. c. node-Mercury
10. n. node-node
11. d. node-Venus
12. l. node-Ascendant
13. j. node-Pluto
14. h. node-Uranus

# Bibliography

Cavendish, Richard, Editor, *Man, Myth & Magic*, New York: Marshall Cavendish Corporation, Vol. 1, 1970

Levinson, Daniel J., et al. *The Seasons of a Man's Life*, New York: Ballantine Books, a division of Random House, 1978.

Negus, Kenneth G. "The Moon's Nodes, The Nineteen-Year Transit Cycle and the U.S. Presidents." The *Journal of the Astrological Society of Princeton*, N. J., Inc., Issue Number 2.

Urdang, Laurence, Editor in Chief. *The Random House Dictionary of the English Language*, New York: Random House, 1968

West, John Anthony and Jan Gerhard Toonder. *The Case for Astrology*, New York: Coward-McCann, Inc., 1970.

# Index

## A

Age Thirty Transition 30, 40, 110
air 121
alchemy 5, 25-26, 34-35, 45-47, 59-60, 74-75, 91-92, 104
    Jupiter 34-35
    Mars 25-26
    Neptune 74-75
    Pluto 91-92
    Saturn 45-47
Allied forces 8
Aquarius 118, 120-121
angles (of the horoscope) 14-15, 17-19
Ascendant 18, 25, 31, 45, 58, 72, 81, 89, 102-103
    conjunct Jupiter 31
    conjunct Neptune 72
    conjunct nodes 102-103
    conjunct Pluto 81, 89
    conjunct Saturn 45
    conjunct Uranus 58
    in solar return 18
aspects 3-4, 18, 20.28, 36, 48, 61, 75, 93, 106, 128
    applying 4
    between eclipse charts and natal planets 20-21
    in solar returns 14-15
    Jupiter transits 30-33
    major configurations (in solar returns) 15, 18
    Mars transits 22-25
    Mercury transits 13-14
    Moon transits 11-12
    Neptune transits 65-74
    node transits 96-103
    Saturn transits 39-45
    separating 4
    Sun transits 13-14
    Uranus transits 51-59
    Venus transits 13-14
    See also individual aspects, e.g. conjunction

Astro Computing Services 129, 132
astrology, validity 1

## B

Becoming One's Own Man 31, 40, 52, 110
benefic 30, 39

## C

calendar transits 132, 134
Cancer 16, 70, 120,
Capricorn 16, 121
cardinal 17, 121
cause-and-effect 1
Childhood and Adolescence 7
conjunctions 4, 8-9, 30-31, 33, 51, 60, 79, 86, 89-90
    Jupiter-Ascendant 33
    Jupiter-Pluto 86
    Mars-Saturn 13
    Pluto-Ascendant 89-90
    Saturn-Neptune 8, 51, 79
    Saturn-Pluto 8-9
    Uranus-MC 15
    Uranus-Pluto 9
    Uranus-Saturn 60

## D

Descendant 18

# E

Early Adult Transition 6-7, 30, 40, 52. 108, 109
Early Adulthood (life stage) 7, 111
earth 120
eclipses 3, **19-21**
    houses 20
    locality 19
elements 120-121
    See also individual elements, e. g. air
Entering the Adult World 30, 109, 111, 128
Entering Middle Adulthood 31
ephemerides, graphic 129-133
exercises (quizzes) 27-28, 35-37, 47-49, 60-62, 75-77, 92-94, 104-106
    answers 135-138
    Jupiter 35-37
    Mars 27-28
    Neptune 75-77
    nodes 104-106
    Uranus 60-63
    Pluto 92-94
    planets in aspect 28, 36-37, 48-49, 61-62, 76-77, 93-94, 106
    planets in houses 27, 35-36, 47-48, 60-61, 75-76, 92-93, 104-105
    Pluto 92-94
    Saturn 47-49
    Uranus 60-62

# F

fatalism 1-2
fire 120
fixed 17, 121-122
free will 1-2
Full moon 19

# G

Gemini 121
gender roles 65
Germany 8
graphic ephemeris 129-133

# H

houses 2-3, 16, 27, 35, 47, 60, 75, 92, 102-103, 114-117
    angular 18
    Jupiter transits 32, 36, 116
    Mars transits 23-25, 27-28, 116
    Mercury transits 13-14, 117
    Moon transits 12, 117
    Neptune transits 64-76, 115
    node transits 103-105, 114-117
    Pluto 84-93, 115
    Saturn transits 45, 47, 116
    Sun transits 14, 117
    Uranus transits 60, 115-116
    Venus transits 14, 117
    1st 17-18, 25
    2nd 112-113
    3rd 3, 17, 87, 103
    4th 16, 18, 69
    5th 87, 103
    6th 73, 81
    7th 15-19, 51, 87, 102, 123
    8th 118
    9th 19-20, 103
    10th 18, 102, 125

# I

I Ching 68, 74
IC 18, 103

# J

Jupiter 3, 4, 6-7, 9, 16, 19, 24, **29-37**, 43, 56, 69, 86, 99, 104, 107-111, 116
aspects and life cycle 29-30, 107-111
exercises (quizzes) 35-37
in houses 33, 35, 116
Jupiter cycle 29-30, 43, 107-111
transits (to other planets) 30-33

# L

Late Adult Transition 7
Late Adulthood (life stage) 7
Leo 8-9, 45, 70
Levinson, Daniel J. 6-7, 30, 37-41, 52, 65-66, 107-114, 128
Libra 8-9, 121
life cycle 6-8, 107-119
    See *also* life stages
life stages
    Becoming One's Own Man 31, 40, 52, 110
    Childhood and Adolescence 7
    Early Adulthood 7, 30
    Entering the Adult World 30, 108, 109, 111, 128
    Entering Middle Adulthood 31
    Late Adulthood 7
    Settling Down 31, 41, 52, 110
Life transitions
    Age Thirty Transition 30, 40, 110
    Early Adult Transition 7, 40, 52, 108, 109
    Late Adult Transition 7
    Mid-life Transition 7, 40-41, 52, 65, 111-113
lunar returns 14
lunations 3, 19-21
locality 19

# M

major configurations
    See aspects
    See T-squares
malefic 38
Mars 3, 13, 16, **22-28**, 32, 43, 55, 69, 81, **85-86**, 99, 116-117, 123
    alchemy 25-26
    exercises (quizzes) 27-28
    in houses 25, 27, 116-117
    Mars cycles 22-23
    transits (to other planets) 22-25
MC, Midheaven 18, 25, 33, 44-45, 58, 71-72, 88-89, 102-103
mentor 31, 110
Mercury 11, **13-14**, 22, 32, 42-43, 54-55, 67-68, 84-85, 98-99, 117
    in houses 14, 117
    movement (speed) 13
Mid-life Transition 7, 40-41, 52, 65, 111-113
modes 17, 120-123
    in solar return 17
    See also individual modes, e. g., fixed
*modus operandi* ( native) 9-10, 117, 122
Moon 3, **11-12**, 14-16, 18-20, 23, 25, 31, 42, 54, 67, 84, 98, 117, 120-121

in houses 12, 117
movement (speed) 11
solar return houses 16
mutable 17, 122

# N

Negus, Kenneth 96
Neptune 2, 3, 6-8, 17, 24, 32-33, 44, 51, 55-56, **63-77**, 79, 88, 101, 112-115
    alchemy 74-75
        exercises (quizzes) 75-77
    in houses 72-73, 75, 115
Neptune cycle 63, 111-114
    transits (to other planets) 64-74
New Moon 19
nodes 17, 25, 33, 44, 57, 71, 88, **95-104**, 114-117
alchemy 104
    exercises (quizzes) 104-106
    in houses 104, 114-117
    nodal cycle 95-96
    transits (to planets) 96-103

# O

oppositions 4, 6, 30, 41, 52-53, 68, 70, 103, 109
orb 4

# P

Part of Fortune 17, 25, 33, 44, 57-58, 71, 88-89, 101-102
planets 28, 36, 48, 61, 95,
    inner 3
    outer 3, 6
    outer cycles of 6-10, 108-114
Pluto 3, 5-6, 8-9, 17, 24, 33, 57, 70-71, **78-94**, 101, 115

alchemy 91-92
exercises (quizzes) 92-94
    in houses 89, 90, 115
    Pluto cycle 78
transits (to other planets) 81-89
polarities 65-66
power 8-9, 70, 79, 83, 87, 89-90
presidents, U.S. 96-97

# Q

quincunxes 4
quizzes (see exercises)

# R

retrograde 4, 14, 63, 78, 126
returns, lunar **14**
Returns, Saturn 5, 40-41, 110, 114
returns, solar 14-19
    locality 15
    major configurations 18-19
    modes on angles 17
    planets conjunct angles 14-15, 18
planets in houses 14-15, 16-17
signs on angles 17-18

# S

Saturn 2, 3, 6-9, 13-16, 19, 24, 32 **38-49**, 51, 56, 59, 64, 69. 79, 83, 86-87, 93, 100 116
alchemy 44-46
exercises (quizzes) 47-49
in houses 44, 47, 114
Saturn cycle 38-41, 45, 107-114
Saturn Return 6, 7, 40, 110, 114
transits (to others planets) 40-44

science and astrology 1
Scorpio 46, 122
*The Seasons of a Man's Life* 6, 30, 39, 41, 52, 65, 108, 109, 110
   See also Levinson, Daniel J.
Settling Down (life stage) 31, 40, 52, 110
sextiles 4, 88, 124
signs 18, 118, 124
   on Ascendant in solar return 17-18
solar returns 14-16, 18-21
squares 3-4, 7, 30, 40-42, 52, 55, 66-67, 82-84, 87, 108-119, 128
stellium 46, 118, 120
Sun 1, 4, 11, **13-21**, 23, 26, 31, 41, 54, 67, 83, 98, 117-118, 123-126, 129
houses 14, 117
movement (speed) 13-14

## T

transit calendar 132, 134
transit categories 6-10
transit sheets, monthly 124-126
transit sheets, six month 126-129
transits, as information 1-10
transits, customizing 117-124
transits, directing 1-5 transits, internal and external 2-4
transits, positive and negative options for 1-2, 10
trine 4, 123
T-squares 19, 123

## U

Uranus 3, 7-8, 14-17, 19, 24, 32, 43-44, **50-62**, 70, 79, 87, 88, 93, 100-101, 104, 108-116, 123, 128
alchemy 59-60
exercises (quizzes) 60-62
in houses 58-59, 60, 116
returns 53-54
transits (to other planets) 54-58
Uranus cycle 40, 50-53, 108-114, 128

## V

Venus 9, 11, **13-14**, 16, 24, 32, 43, 55, 68, 85, 99, 117, 126
   in houses 14, 117
   movement (speed) 13
Virgo 18

## W

water 121
World War II 8

## Y

Yale 6
yod 88

# About the Author

## Joan Negus
### 1930-1997

Joan Negus, who was born July 30, 1930 at 6:04 am EDT in Trenton, New Jersey, is remembered as one of astrology's finest educators. She and Joanna Shannon, as Co-Directors of Education for National Council for Geocosmic Research, Inc., were the prime movers in the development of the organization's comprehensive four level education and testing program. Joan's devoted service to astrological education was honored at the 1995 United Astrology Congress with the prestigious Regulus Award for Education. Joan and her husband Ken together won the Matrix Pioneer Award for their leadership in setting high professional standards in the field of astrology.

Joan also helped found the Astrological Society of Princeton (New Jersey). She and Ken, through their organization work and also through many gatherings in in their home, provide a constant core of energy for the astrological community—workshops, classes and spirited group discussions. Joan also practiced as an astrological consultant and had a large clientele. She was the author of five books, providing a thorough and systematic approach to both teaching and learning astrology that has benefited many, both students and teachers—and will continue to do so.

# Also by Joan Negus

*Basic Astrology:
A Guide for Teachers and Students*

&

*Basic Astrology:
A Workbook for Students*

*Cosmic Combinations*

*The Book of Uranus*

*Interpreting Composite and
Relationship Charts*

*Astro-Alchemy:
Making the Most of Your Transits*

---

For your **One-Year Graphic Ephemeris**, as shown on page 133 or your **Calendar Transits**, as shown on page 134, or for a wide variety of other charts, calculations and reports, contact

## Astro Computing Services

Orders are normally delivered same day Monday through Friday in PDF format by email, or are printed and mailed, as you request.

**Order online:** secure shopping cart—*www.astrocom.com*
**By mail:** 334-A Calef Highway, Epping, NH 03042
**By phone:** 24-hour toll free message line: 866-953-8458
**Our office phone:** 603-734-4300, Monday-Friday 9am-4:30 pm
Receive our monthly newsletter with special offers. Opt in!

# Personalized Astrology Lessons

The only correspondence course that teaches you astrology with examples from your personal horoscope.

# P.A.L.S

**PAL LESSON TOPICS**

1. Introducing Astrology: Planets, Signs and Houses 2.The Glyphs and More on Planets & Signs 3. The Astrological Alphabet 4. More on Planets in Signs 5. More on Planets in Houses 6. Elements and Qualities 7. Rulers—Natural & Actual 8. Introducing Aspects 9. Learning to Spot Aspects & Meanings 10. Interpreting Planetary Aspects 11. Integrating Houses & Signs with Planetary Aspects 12. More on Synthesizing 13. More Themes 14. Spotting Repeated Themes 15. Odds Ends (retrogrades, stations, house systems, interceptions & more) 16. Odds & Ends Continued (exaltaton, fall, detriment, rulers & more) 17. Odds & Ends Continued (East Point, Vertex, Moon's Nodes & more) 18. Identifying Life Areas in the Horoscope 19. Analyzing Basic Identity 20. Analyzing Career 21. Analyzing Relationships 22. Analyzing Mind & Communication 23.Analyzing Parents 24.Mother & Father 25. Analyzing Children & Creativity 26. Analyzing Financial Prospects 27. Analyzing Beliefs & Values 28] Analyzing Sensuality & Sexuality 29. Karmic Lessons, 30. Future Potentials 31. Health Options in the Horoscope. 32. Analyzing Health Options.

These lessons offer you an opportunity to master the age-old discipline of astrology, and will empower you to look deeper into all the issues of your life. Enhance your self-esteem and discover your highest potential. Maritha Pottenger created these lessons to help those curious beginners to understand all the tools that astrology has to offer. With an activist approach to astrology, Maritha shows you how to create your future and the life you want! Reinforce these lessons with actual homework assignments that test your knowledge! Please specify Lesson Numbers when ordering.

All 32 Lessons on 1 chart with Notebook
PAL ALL-BOW1..................................................$99.95
One Lesson PAL-BOW1........................................5.95

Any 6 Lessons on 1 chart PAL6-BOW1.....$24.95
Notebook only NB-BOW1................................8.95

Prices subject to change. Shipping & handling will be added.

**ASTRO COMPUTING SERVICES**
Starcrafts LLC, PO Box 446, Exeter, NH 03833

# www.astrocom.com

# The American Ephemeris Series

All of the standard-setting reference works

by Neil F. Michelsen and Rique Pottenger

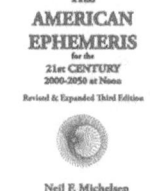  

are now available in newly updated editions

**Starcrafts Publishing & ACS Publications**
imprints of Starcrafts LLC, 334-A Calef Hwy, Epping, NH 03042
603-734-4300 • www.astrocom.com

## Also from Starcrafts LLC
**Imprints:** *Starcrafts Publishing, ACS Publications*

*All About Astrology*, a series of booklets by various authors

*The American Atlas, Expanded 5th Edition*, Thomas G. Shanks

*The American Ephemeris for the 21st Century [at Noon or at Midnight] 2000-2050, Rev. 2nd Ed.*, Neil F. Michelsen, Revisions by Rique Pottenger

*The American Heliocentric Ephemeris 2001-2050*, Neil F. Michelsen

*The American Sidereal Ephemeris 2001-2025*, Neil F. Michelsen

*The Asteroid Ephemeris 1900-2050*, Rique Pottenger with Neil F. Michelsen

*Astrology for the Light Side of the Brain*, Kim Rogers-Gallagher)

*Astrology for the Light Side of the Future*, Kim Rogers-Gallagher)

*Astrology: the Next Step*, Maritha Pottenger

*Astrology and Weight Control*, Beverly Ann Flynn

*The Book of Jupiter*, Marilyn Waram

*Dial Detective, Revised Second Edition*, Maria Kay Simms

*Easy Astrology Guide*, Maritha Pottenger

*Easy Tarot Guide*, Marcia Masino

*Future Signs*, Maria Kay Simms

*The International Atlas, Revised 6th Edition*, Thomas G. Shanks & Rique Pottenger

*The Michelsen Book of Tables*, Neil F. Michelsen

*Moon Tides, Soul Passages*, Maria Kay Simms, with software CD by Rique Pottenger

*The New American Ephemeris for the 20th Century, 1900-2000, at Midnight Michelsen Memorial Edition*, Rique Pottenger, based on Michelsen

*The New American Ephemeris for the 20th Century, 1900-2000, at Noon Michelsen Memorial Edition*, Rique Pottenger, based on Michelsen

*The New American Ephemeris for the 21st Century, 2000-2100 at Midnight Michelsen Memorial Edition*, Rique Pottenger, based on Michelsen

*The New American Ephemeris for the 21st Century, 2007-2020: Longitude, Declination, Latitude & Daily Aspectarian*, Rique Pottenger, based on Michelsen

*The New American Midpoint Ephemeris 2007-2020*, Rique Pottenger, based on Michelsen

*The Only Way to Learn Astrology, Volumes. 1-6 series* Marion D. March & Joan McEvers

*Past Lives, Future Choices*, Maritha Pottenger

*Pathways to Success*, Gayle Geffner

*Planetary Heredity*, Michel Gauquelin

*Planets on the Move*, Maritha Pottenger and Zipporah Dobyns, Ph.D.

*Psychology of the Planets*, Francoise Gauquelin

*Tables of Planetary Phenomena, Third Edition*, Neil F. Michelsen

*Unveiling Your Future*, Maritha Pottenger and Zipporah Dobyns, Ph.D.

*Yankee Doodle Discord: A Walk with Eris through USA History*, Thomas Canfield

*Your Magical Child*, Maria Kay Simms

*Your Starway to Love*, Maritha Pottenger

www.ingramcontent.com/pod-product-compliance
Lightning Source LLC
LaVergne TN
LVHW051837080426
835512LV00018B/2923